OXFORD MEDICAL PUBLICATIONS

A Practical Guide to Chorion Villus Sampling

A Practical Guide to Chorion Villus Sampling

David T. Y. Liu

Senior Lecturer and Honorary Consultant Obstetrician and Gynaecologist, Department of Obstetrics and Gynaecology, City Hospital, Nottingham

Oxford New York Tokyo
OXFORD UNIVERSITY PRESS
1991

Oxford University Press, Walton Street, Oxford OX2 6DP
Oxford New York Toronto
Delhi Bombay Calcutta Madras Karachi
Petaling Jaya Singapore Hong Kong Tokyo
Nairobi Dar es Salaam Cape Town
Melbourne Auckland
and associated companies in
Berlin Ibadan

Oxford is a trade mark of Oxford University Press

Published in the United States
by Oxford University Press, New York

British Library Cataloguing in Publication Data
A practical guide to chorion villus sampling.
1. Antenatal medicine. Diagnosis
I. Liu, D. T. Y. (David Tek-Yung)
618.22

ISBN 0-19-262006-1
ISBN 0-19-262005-3 (pbk)

Library of Congress Cataloging in Publication Data
A practical guide to chorion villus sampling/[edited by] David T.Y. Liu.
p. cm.
Includes index.
1. Chorionic villus sampling. I. Liu, D. T. Y.
[DNLM: 1. Chorionic Villi Sampling—methods. 2. Genetic Screening—methods. 3. Prenatal Diagnosis—methods. WQ 209 P895]
RG628.3.C48P73 1991 90-7937
618.3′2042—dc20

ISBN 0-19-262006-1 ISBN 0-19-262005-3 (pbk.)

Typeset by Cotswold Typesetting Ltd, Cheltenham
Printed in Great Britain by
Biddles Ltd., Guildford &
King's Lynn

Contents

Contributors

Dr Pat Cooke and Mrs Gerardine Turnbull-Ross, South Trent Cytogenetic Service, Hucknall Road, Nottingham NG5 1PB

Mr David T. Y. Liu, Senior Lecturer/Consultant, Department of Obstetrics and Gynaecology, City Hospital, Hucknall Road, Nottingham NG5 1PB

Mr D. J. Maxwell, Consultant Senior Lecturer, Department of Obstetrics and Gynaecology, Guy's Hospital, Floor 2, New Guy's House, London Bridge, London SE1 9RT

Dr Roger Quaife, formerly Principal Scientist, Molecular Genetics Unit, City Hospital, Hucknall Road, Nottingham NG5 1PB

Miss Rosalyn Richardson, Project Manager—Clinical Information Systems, Forest House, Berkeley Avenue, Nottingham NG3 5AF

Dr Denise E. Rooney, Cytogenetics Unit, St Mary's Hospital Medical School, Norfolk Place, London W2 1PG

Dr P. Twining, Consultant Radiologist, University Hospital, Queen's Medical Centre, Nottingham NG7 2UH

Dr I. D. Young, Senior Lecturer, Department of Child Health, Leicester Royal Infirmary, Leicester LE1 5WW

1 Introduction and historical perspectives

David T. Y. Liu

1.1 Rationale for current interest

There are several major reasons why prenatal diagnosis of congenital abnormalities in the fetus is attracting increasing interest. The fall in perinatal mortality over the last three decades to levels below ten per thousand means that congenital abnormalities now account for a substantial (four per thousand) proportion of contemporary perinatal wastage (Butler and Bonham 1961–3; Social Services Committee 1980). This perinatal loss can be improved by identification and early removal of fetuses with anomalies that are not compatible with life.

In the developed countries the expectations of society in a more materialistic and demanding world have also contributed to this increase in interest. The pursuit of perfection is paramount, and there is much less incentive or willingness to accept deviation from normality. In the developing countries, or where supportive health services are not available, the cost implications of a child with inherited problems can be prohibitive.

Prenatal diagnosis, however, is not a new concept; albeit common usage has relegated it from consideration as such because of its application mainly within the third trimester. Examples range from the time-honoured detection of heart-rate with a Pinard stethoscope, to external cardiotocography or ultrasound scanning. Options are limited when fetal problems are detected at this late stage of pregnancy; delivery can be expedited by induction of labour, or, where fetal jeopardy is severe, elective Caesarean section may prove prudent.

Investigations of the fetus in the second trimester are traditionally equated with the implications of prenatal diagnosis. Increasing sophistication of ultrasound machines and techniques, together with an accumulating wealth of experience, allow identification of many fetal structural abnormalities. Amniocentesis provided an invaluable service for several decades, and remains the only means for assessing the biochemical status of the amniotic fluid. Access to amniocytes for cytoculture allows examination of fetal chromosome configuration or provides fetal deoxyribonucleic acid (DNA) for gene probing (Schneck et al. 1970; Chan et al. 1984).

Amniocentesis is not without complications. There is an overall 7 per cent (range 0–14 per cent) failure to obtain amniotic fluid, even when ultrasound is used (MRC study 1978). Feto-maternal haemorrhage, and thus raised maternal serum alpha-fetoprotein, is observed in up to 10 per cent of cases, despite no fetal or placental trauma (Stephenson and Weaver 1981). An intervention fetal loss-rate of 1 per cent (range 0.5–1.5 per cent) on top of the spontaneous miscarriage rate of 1.0–1.7 per cent is well recognized. This loss-rate is increased when the operator is inexperienced, and doubled when there is placental damage. A morbidity rate of 1.5 per cent, made up of preterm labour, neonatal respiratory difficulties, and antepartum haemorrhage is observed (MRC study 1978). Repeat amniocentesis is required in some 1.7–4.9 per cent of cases. When this happens, results may not be available until after 20 weeks of pregnancy, when therapeutic abortion can present ethical difficulties. About 0.9 per cent of women will choose not to repeat the procedure.

Cytoculture failure—and hence inability to provide a diagnosis—occurs with 3–5 per cent of the amniotic samples. This is particularly likely when the amniotic fluid is bloodstained (4.5–13 per cent). Wrong results as a consequence of maternal cell contamination is reported in 0.5 per cent of cases (Gosden 1987).

The need for a more comprehensive diagnostic capability particularly for the haemoglobinopathies, which affect a considerable number of people in developing countries, encourages attempts to gain direct access to fetal blood. Placentocentesis, which involves repeated placental puncture and aspiration to collect a mixture of amniotic fluid and fetal and maternal blood was suggested (Golbus *et al.* 1976). A minimum of some 5 per cent of fetal cells are required for haematological analysis (Fairweather *et al.* 1980). The potential of this crude diagnostic approach is limited by anxieties concerning miscarriage, contamination by maternal components, and the paucity of fetal cells (Fairweather *et al.* 1980).

Fetoscopy is the technique whereby an endoscope, usually 1.7 mm in diameter, is introduced under aseptic conditions into the mid-trimester uterus at a site predetermined by ultrasound examination. A fibre-optic light-source provides illumination, permitting direct visualization of the topography of the fetus. A fine (27-gauge) needle can also be introduced through a side channel to access cord blood under direct vision. The ability to obtain fetal blood allows ready and prompt examination of a vast array of problems, thus immediately expanding the diagnostic scope.

Despite the undoubted advantages for diagnosis, fetoscopy carries a 5 per cent (3–10 per cent) risk of miscarriage. Leakage of amniotic fluid associated with the sizeable puncture contributes to the reported 10–15 per cent preterm labour rate. The intricacy and risk associated with this procedure means that fetoscopy should be restricted to specialized departments and performed only by skilled operators. Although the technique was popular in the early 1970s (Hobbin *et al.* 1974; Rodeck and Campbell 1978), and stimulated much interest in prenatal diagnosis, the high complication-rate led investigators to explore simpler and safer means to achieve the same result.

Daffos *et al.* (1983) and Nicolaides *et al.* (1986) introduced and promoted the technique of cordocentesis. This out-patient procedure involves ultrasound-guided passage of a 20-gauge needle through the placenta (anterior presentation) or through the amniotic fluid (posterior presentation) into the umbilical cord to obtain fetal blood. Fetal loss, due principally to chorio-amnioitis or haemorrhage, is between 1.5 and 2.0 per cent. Although the safety of the procedure approximates to that for amniocentesis, it nevertheless requires a certain degree of skill, and is not as readily incorporated into the obstetrician's surgical repertoire as amniocentesis.

Common to all mid-trimester diagnostic procedures is the advanced stage of gestation at which results become available. Apart from an anxious wait for diagnosis, this delay can cause religious, moral, and ethical dilemmas for many couples in need of prenatal assessment: many cannot or will not entertain interference after the first trimester. On the other hand, if a couple decides to terminate the pregnancy, mid-trimester therapeutic abortion is costly, both in terms of emotion and of economy, and subjects the woman to increased morbidity (3–5 per cent), with an eight- to tenfold increase in mortality. These aspects must be taken into consideration when couples or families are counselled.

Prenatal diagnosis achieved within the first trimester will overcome many of the above problems. Anxiety for at-risk patients is reduced, and the pregnancies can be kept private between the doctors and the couples concerned until normality is confirmed. If abnormality is detected and resolution is considered appropriate this can be accomplished safely and expeditiously by first-trimester therapeutic abortion on a day-case basis. This desirable situation can be achieved if chorionic villus sampling is used for prenatal diagnosis.

1.2 Embryology

Trophoblast derived from the same cell lineage as the fetus appears at around the 64-cell stage. The first-trimester embryo is enveloped by this tissue, which at this stage exceeds the total volume of the fetus. At 8 weeks the trophoblast weighs 5 grams, compared to the 0.5 grams of the fetus. At 12 weeks the weights are 80 grams and 25 grams respectively (Hamilton and Hamilton 1977). Once the embryo becomes embedded in the decidua, strands of chorionic villi migrate into the decidua basalis to form the chorion frondosum, or primitive placenta. As happens in all primates, primitive plasmodial trophoblast differentiates early in pregnancy into syncytiotrophoblast and cytotrophoblast. Twelve days after implantation cytotrophoblast begins to form generations of primary, secondary, and tertiary chorionic villi (Hamilton and Hamilton 1977). Fetal vessels in the villi are observed towards the end of the third week. Villi penetrating into maternal blood-vessels create a haemochorial placenta, and allow emboliz-ation of trophoblast into the maternal circulation.

Compared with the chorion frondosum, chorionic villi at extra-placental sites tend to be shorter and less branched, and attenuate by 14 weeks, when the gestation sac approximates with the inner wall of the uterine cavity. Removal of a small sample of this villus material should not therefore jeopardize subsequent fetal development. Villus tissue is also actively dividing, and thus many mitotic figures are present: removal of an adequate amount for karyotyping by direct-preparation techniques will obviate the need for cytoculture. The risk of cytoculture failure can be eliminated, and results can be available within days, if not hours.

Good-quality deoxyribonucleic acid (DNA) is extracted from villus material (Chapter 3). The increasing number of gene probes available for the diagnosis of an expanding list of inheritable conditions means that a ready market exists to exploit the utility of villus tissue. In the 1990s and be-yond there is every potential for the concept of first-trimester diagnosis to achieve practical reality.

Despite this welcome situation, not all congenital abnormalities are currently amenable to diagnosis by examination of chorionic villi. Patients should be made aware that structural malformations, such as neural tube defects, may not become obvious until pregnancy is more advanced.

1.3 Historical perspectives

If we accept that villus tissue is useful for prenatal diagnosis, the next question must be the mode of access to this material. There are essentially three approaches.

The human haemochorial placenta allows embolization of syncytiotrophoblast into the maternal circulation. Although most of this material is sieved out by passage through the lungs, small fragments reach the arterial circulation, and can be collected by venepuncture. Theoretically this would be an ideal approach, since there is no risk whatsoever to the fetus. Raafat *et al.* (1975) were the first to consider collection of trophoblast from maternal venous blood for prenatal diagnosis. They were not successful because the number of trophoblast 'cells' in the peripheral circulation is limited, and since this deported tissue is usually syncytiotrophoblast there is little tendency to undergo mitosis or respond to cytoculture. In the early 1980s the Placental Immunology Group from Liverpool, England, succeeded in raising monoclonal antibodies specific for trophoblast. When a flow cytometry (or cell-sorter) (FACS) machine is incorporated to identify fetal cells and enrich their capture, a sufficient quantity may be obtained for prenatal diagnosis (Covone *et al.* 1984). Parks and Herzenberg (1982) used Quinacrine mustard staining of Y chromatin to confirm the feasibility of this approach by identifying the presence of male fetal cells. If results are confirmed (Rushton 1984), gene amplification techniques could feasibly be applied to the small amount of DNA present to achieve diagnosis. This concept is not as yet a practical proposition; but the challenge undoubtedly encourages the hope of realizing this goal in the near future.

Alvarez (1964) first reported use in the transabdominal approach to obtain chorionic material by fine-needle aspiration for the diagnosis of hydatidiform mole. No complication was observed when this procedure was performed at 10–12 weeks' gestation in 50 patients. Aladjem (1969) attempted diagnosis of placental pathology in 215 third-trimester patients by transabdominal villus sampling. Again no fetal or maternal complication was reported. Smidt-Jensen and Hahnemann (1984), however, were accredited with being the first to adopt transabdominal chorion villus sampling for first-trimester prenatal diagnosis. The feasibility of the approach was confirmed by Maxwell *et al.* (1986).

The cervical canal offers a natural route into the uterine cavity, and one which is familiar to all obstetricians. Moreover, the canal is some 2–3 mm in diameter, and can accommodate with ease a variety of implements for the collection of villus material. It is therefore not surprising to find that transcervical chorion villus sampling is one of the most explored approaches, and currently the most popular.

In the first trimester, before the gestation sac approximates with the uterine wall, trophoblast cells exfoliate and drop on to the operculum at the level of the internal cervical os. Cotton wool swabs (Shettles 1971) and specially designed instruments (Rhine and Milunsky 1979) have been introduced through the cervical canal to collect these exfoliated cells for chromosome analysis to determine fetal sex. Major drawbacks include the paucity of the total number of cells obtained, particularly those which are viable, and inability to achieve certainty as to the origin of the cells. Fetal chromosomes can only be demonstrated in half (49 per cent) of the studied patients (Bobrow and Lewis 1971; Rhine and Milunksy 1979).

Hahnemann and Mohr (1969) first attempted transcervical biopsy of chorionic villi using a hysteroscope of their own design which is 5–7 mm in diameter. Chorionic villus material was sucked into a side-opening of the endoscope and cut with a knife built into the instrument. Hahnemann and Mohr demonstrated that chorionic tissue can be obtained in the first trimester for successful karyotyping. Kullander and Sandahl (1973) confirmed their findings. Under direct endoscopic vision, chorionic material was obtained from gestations of between 8 and 20 weeks for successful karyotyping. A fine biopsy forceps was passed down the barrel of their 5 mm-diameter instrument to collect the diagnostic material. All the above-described endoscopes used were, however, sizeable implements, and bleeding was commonly encountered. In Hahnemann's (1974) series of studies, a clear view was only achieved 50 per cent of the time. Gustavii (1984) improved visualization by instillation of a physiological solution into the uterine cavity throughout the endoscopic procedure. Successful karyotyping in ongoing pregnancies was achieved. The size of the endoscopes used, the need for cervical dilatation, and instillation of fluid all proved too time-consuming, risky for the pregnancy, and cumbersome to attract support for this approach.

In the 1970s the Chinese reported success with blind transcervical chorion sampling (Tietung Hospital, Anshan 1975). A rigid 3 mm metal cannula with a blunt end was inserted, and the likely placental site was localized by feel. Chorionic villus material was collected by aspiration using a 5 mm syringe. Prenatal sex-determination was achieved in 94 per cent of patients sampled at between 6 and 14 weeks' gestation. Miscarriage in ongoing pregnancies was 6 per cent. Currently, China is the only country where children born after villus sampling can be followed from diagnosis to adolescence.

However, two independent studies (Liu *et al.* 1983; Horwell *et al.* 1983) showed that chorionic material was obtained in only a third of blind sampling attempts. Perforation of the gestation sac, and thus inevitable miscarriage, can occur. Furthermore, presence of anembryonic pregnancies, missed abortions, and multiple gestation will complicate results. Kazy *et al.* (1982) overcame these caveats by introducing ultrasound screening and maintenance of scanning to direct the biopsy implement to the chosen placental site. Apart from improving the safety of the procedure, these authors also showed that chorionic tissue has potential for use in the diagnosis of genetic disease and inherited metabolic disorders. Their work, together with the prevailing climate of diagnostic sophistication and support from developments in the paramedical sciences, contributed much to encourage redirection of interest towards chorion villus sampling for prenatal screening.

1.4 Consequences

It is commonly observed that new procedures introduced into obstetric practice often incorporate consequences which infringe upon our notions of moral, social, and medical propriety. Chorion villus sampling is one such procedure. What could be the results of opening this particular Pandora's box?

The ease of the procedure and the ever-increasing ability to gene probe for an expanding catalogue of inheritable problems may tempt us to the application of this technique to a wider spectrum of pregnant women. Is mass screening conceivable? In countries where non-genetic sexing of the conceptus is permitted, villus sampling has already been used for the preferential selection of a particular sex. If this practice is allowed to continue, distortions in the balance of our society may be anticipated. Many prenatal diagnostic departments correctly refrain from revealing the sex of the fetus unless there is the possibility of a sex-linked genetic problem.

The escalation of interest in early prenatal screening initiated by villus sampling could lead to fulfilment of its potential for preconceptual screening, which will benefit couples who believe that therapeutic abortion cannot be justified. Techniques for fertilization *in utero* can also be applied to produce conception outside the uterine cavity. One or more blastomerases can be removed from the pre-implantation conceptus at the 4–16 cell stage for diagnostic purposes. The remaining cells are cryo-preserved, and implanted only when normality is confirmed. It has already been shown in the mouse that pre-implantation diagnosis can succeed; its adaptation to human usage need not be far behind (Penketh and McLaren 1987). On a more futuristic note, this same process could form the basis for gene therapy of monogenetic defects in the conceptus, when implantation would be carried out only after correction or deletion of the genetic defect (McLaren 1984).

1.5 Aims of this book

One of the aims of this book is to introduce the reader to the historical perspective of chorion villus sampling, and to its relevance to current obstetric practice. An attempt is also made to indicate the possible direction of development of villus sampling in the near future. Sufficient background information is given to allow obstetricians intending to adopt the technique to provide the comprehensive counselling expected of them. Obstetricians should become familiar both with the transabdominal and the transcervical approaches. Each has its own advantages and caveats. Detailed step-by-step instruction is provided for this purpose. Information is also available to guide the reader from the training stage to the implementation of the technique to provide a service.

Whether villus sampling will fulfil all the expectations conceived for it remains to be seen. Only the court of history can adjudicate the contribution of villus samplings to obstetric progress. The editor, a proponent of the technique, is acutely aware that he will be answerable to this same court.

References

Aladjem, S. (1969). Fetal assessment through biopsy of the placenta. In *The feto-placental unit* (ed. A. Pecile and C. Finzi), pp. 392–402. Excerpta Medical Foundation, Amsterdam.

Alvarez, H. (1964). Morphology and physiopathology of the human placenta. *Obstetrics and Gynaecology*, **28**, 813–15.

Bobrow, M. and Lewis, B. V. (1971). Unreliability of fetal sexing using cervical material. *Lancet*, **ii**, 486.

Butler, N. R. and Bonham, D. G. (1963). *Perinatal mortality: the first report of the 1958 British Perinatal Mortality Survey.* Livingstone, Edinburgh.

Chan, V., Ghosh, A., Chan, T. K., Wong, V., and Todd, D. (1984). Prenatal diagnosis of homozygous alpha thalassaemia by direct DNA analysis of uncultured amniotic fluid cells. *British Medical Journal*, **288**, 1327–9.

Covone, A. E., Mutton, D., Johnson, P. M., and Adinolfi, M. (1984). Trophoblast cells in peripheral blood from pregnant women. *Lancet*, **ii**, 841–3.

Daffos, F., Cappella-Pavlovsky, M., and Forestier, F. (1983). Fetal blood sampling via the umbilical cord using a needle guided by ultrasound. Report of 66 cases. *Prenatal Diagnosis*, **3**, 277.

Fairweather, D. V. I., Ward, H. R. T., and Modell, B. (1980). Obstetric aspects of mid-trimester fetal blood sampling by needling or fetoscopy. *British Journal of Obstetrics and Gynaecology*, **87**, 87–99.

Golbus, M. S., Kan, Y. W., and Naglich-Craig, M. (1976). Fetal blood sampling in mid-trimester pregnancies. *American Journal of Obstetrics and Gynaecology*, **124**, 653–5.

Gosden, C. (1987). Prenatal karyotyping: amniotic fluid cells or chorion villus samples? In *Chorion villus sampling* (ed. D. T. Y. Liu, E. M. Symonds, and M. S. Golbus), pp. 257–72. Chapman and Hall Medical, London.

Gustavii, B. (1984). Chorionic villi sampling under direct vision. *Clinical Genetics*, **26**, 297–300.

Hahnemann, N. (1974). Early prenatal diagnosis: a study of biopsy techniques and cell culturing from extra-embryonic membranes. *Clinical Genetics*, **6**, 294–306.

Hahnemann, N. and Mohr, J. (1969). Antenatal fetal diagnosis in genetic disease. *Bulletin of European Society of Human Genetics*, **3**, 47–54.

Hamilton, W. J. and Hamilton, D. V. (1977). Development of the human placenta. In *Scientific foundations of obstetrics and gynaecology* (ed. E. E. Philip, J. Barnes, and M. Newton), Heinemann, London.

Hobbin, J. C., Mahoney, M. J., and Goldstein, L. A. (1974). New method of intra-uterine evaluation by the combined use of fetoscopy and ultrasound. *American Journal of Obstetrics and Gynaecology*, **118**, 1069–1972.

Horwell, D. H., Loeffler, F. E., and Coleman, D. V. (1983). Assessment of a trans-cervical technique for chorionic villus biopsy in the first trimester of pregnancy. *British Journal of Obstetrics and Gynaecology*, **90**, 196–8.

Kazy, Z., Rozovsky, I. S., and Bakharev, V. A. (1982). Chorion biopsy in early pregnancy: a method of early prenatal diagnosis of inherited disorders. *Prenatal Diagnosis*, **2**, 38–45.

Kullander, S. and Sandahl, B. (1973). Fetal chromosome analysis after trans-cervical placental biopsies during early pregnancy. *Acta Obstetrica et Gynaecologica Scandinavica*, **52**, 355–9.

Liu, D. T. Y., Mitchell, J., Johnson, J., and Wass, D. M. (1983). Trophoblast sampling by blind transcervical aspiration. *British Journal of Obstetrics and Gynaecology*, **90**, 119–23.

Maxwell, D., Lilford, R., Czepulkowski, B., Heaton, D., and Coleman, D. (1986). Transabdominal chorion villus sampling; development and clinical application. *Lancet*, **i**, 123–6.

McLaren, A. (1984). Prenatal diagnosis before implantation: opportunities and problems. *Prenatal Diagnosis*, **5**, 85–90.

MRC Working Party on amniocentesis (1978). An assessment of the hazards of amniocentesis. *British Journal of Obstetrics and Gynaecology*, **85** (Supplement 2), 1–41.

Nicolaides, K. H., Soothill, P. W., Rodeck, C. H., and Campbell, S. (1986). Ultrasound guided sampling of umbilical cord and placental bood to assess fetal well-being. *Lancet*, **i**, 1065–7.

Parks, D. R. and Herzenberg, L. A. (1982). Fetal cells from maternal blood: their selection and prospects for use in prenatal diagnosis. *Methods in Cell Biology*, **26**, 277–95.

Penketh, R. and McLaren, A. (1987). Prospects for prenatal diagnosis during perimplantation human development. In *Clinical Obstetrics and Gynaecology. Fetal Diagnosis of Genetic Defects* (ed. C. H. Rodeck), **1**, no. 3, pp. 747–64. Bailliere, Tindall.

Raafat, M., Brayton, J. B., and Apgar, V. (1975). A new approach to prenatal diagnosis using trophoblast cells in the maternal blood. *Birth Defects Original*

Article Series XI, **5**, 295–302.

Rhine, S. A. and Milunsky, A. (1979). Utilisation of trophoblast for early prenatal diagnosis. In *Genetic Disorder and the Fetus* (ed. A. Milunsky), pp. 527–39.

Rodeck, C. H. and Campbell, S. (1978). Sampling pure fetal blood by fetoscopy in second trimester of pregnancy. *British Medical Journal*, **2**, 728–30.

Rushton, I. (1984). Trophoblast cells in peripheral blood. *Lancet*, **ii**, 1153–4.

Schneck, L., Valenti, C., Amsterdam, D., Friedland, J., Adachi, M., and Volk, B. W. (1970). Prenatal diagnosis of Tay–Sachs disease. *Lancet*, **i**, 582–3.

Shettles, L. B. (1971). Use of the chromosome in prenatal sex determination. *Nature*, **230**, 52.

Smidt-Jensen, S. and Hahnemann, N. (1984). Transabdominal fine needle biopsy from chorionic villi in the first trimester. *Prenatal Diagnosis*, **4**, 163–9.

Social Services Committee (1980). Prenatal and neonatal mortality. Second report from the Social Services Committee, 1979–80. HMSO, London.

Stephenson, S. R. and Weaver, D. D. (1981). Prenatal diagnosis: a compilation of diagnosed conditions. *American Journal of Obstetrics and Gynaecology*, **141**, 319–43.

Tietung Hospital, Anshan (1975). Fetal sex prediction by sex chromatin of chorionic villi cells during early pregnancy. *Chinese Medical Journal*, **1**, 117–26.

2 Genetic counselling

I. D. Young

2.1 Introduction

The happily coincidental developments of chorion villus sampling and molecular biotechnology have provided radical new opportunities for parents to exercise individual choice when embarking upon one of life's greatest adventures, the creation of a human being. This medical revolution, the pace of which has vastly exceeded even the most optimistic predictions, has far-reaching implications for parents, for society, for medical practice, and not least for a legal profession alert to the lucrative rewards of malpractice litigation. Increasing awareness of the complexities of prenatal diagnosis, particularly those aspects relating to chorion villus sampling, has led to general agreement that it is in the best interests of both patient and doctor if prenatal diagnostic services utilize an integrated approach embracing the collective skills of the obstetrician, the laboratory scientist, and the clinical geneticist (King's Fund Forum 1987). It is with this latter role that this chapter is concerned.

2.2 The spectrum of inherited disease

Traditionally, inherited disorders are classified under three headings—-chromosomal, single gene, and multifactorial. To these has recently been added a fourth category—somatic genetic disease—relating to those disorders, chiefly neoplastic, which are generated by environmentally induced mutations in the genome. Clearly consideration of this fourth group of disorders is outside the remit of this review.

Chromosomal disorders

The normal human being has 46 chromosomes, comprising 22 pairs of autosomes and a single pair of sex chromosomes—XX in the female and XY in the male. During meiosis each pair divides, so that each gamete receives a single (haploid) set. Fertilization results in restoration of the normal (diploid) chromosome constitution in the zygote. Any aberration of the autosomes resulting in loss or gain of chromosome material is likely to lead to mental retardation and physical abnormalities. Alterations in the sex chromosomes are generally less deleterious. Chromosome abnormalities are believed to account for up to 50 per cent of all embryonic and fetal deaths and around 5 per cent of all stillbirths and neonatal deaths, and to occur in approximately 1 in 150 of all new-born infants (Hsu 1986).

Single-gene (Mendelian) disorders

Over four thousand conditions or traits showing single-gene inheritance have been identified (McKusick 1988). A condition which is manifest in the heterozygote, who by definition has only one copy of the abnormal gene, is said to be dominant. A recessive disorder is expressed only in the homozygote, who has two copies of the abnormal gene, having inherited one from each parent. If the abnormal gene is sited on an autosome, then the resulting condition is said to show autosomal dominant or recessive inheritance: if it is on the X chromosome then the term 'sex-linked' is applied.

Autosomal dominant inheritance

If a patient has an autosomal dominant disorder, such as Huntington's chorea or neurofibromatosis, then each child of that individual commences life with a 50 per cent chance of having inherited the abnormal gene. Generally, autosomal dominant disorders affect males and females equally. Estimation of genetic risks can become very difficult if, as is often the case, the disorder in question shows either delayed onset (for example, Huntington's chorea; adult polycystic kidney-disease), or a curious phenomenon known as non-penetrance, in which case the condition may appear to skip a generation (for example, retinoblastoma; tuberose sclerosis).

Autosomal recessive inheritance

Each sibling of a child with an autosomal recessive disorder runs a risk of 1 in 4 of being affected. Usually autosomal recessive disorders affect only members of a single sibship in a family, so that other family relatives are likely to be at very low risk for having an affected child. Exceptions to this general rule may arise firstly if the disorder in question has a high gene-frequency in a particular population (for example, cystic fibrosis in Western Europeans, Tay Sachs disease in Ashkenazi Jews, thalassaemia in Mediterranean and certain Asian groups, and sickle-cell disease in Afro-Caribbeans) and secondly if members of the family have married blood-relatives.

Sex-linked recessive inheritance

Sex-linked recessive disorders, such as Duchenne muscular dystrophy and haemophilia, generally affect only males, with a few rare exceptions, such as 'females' with testicular feminization, patients with Turner's syndrome, and women who carry X-autosome translocations, resulting in non-random X-chromosome inactivation. Each time a female carrier of an X-linked recessive disorder has a child there is 1 chance in 2 that a son will be affected or that a daughter will be a carrier. When an affected male has children, all sons will inherit his Y chromosome, and will thus be unaffected, whereas all daughters will inherit his X chromosome, bearing the abnormal gene, and thus will necessarily be carriers.

Multifactorial disorders

Many common conditions show a clear familial tendency that does not conform to any recognized mode of single-gene inheritance. It is postulated that these disorders result from the interaction between an underlying familial predisposition that is controlled by many genes at different *loci* (a 'polygenic' predisposition) and adverse environmental factors, which are often poorly understood. Common examples include malformations, such as non-syndromal cleft lip/palate, congenital heart-disease, and neural tube defects, and also some chronic disorders of adult life, such as schizophrenia and epilepsy. Although vigorous efforts are being made, with considerable success, to isolate genes which contribute to familial predisposition for these disorders, none is yet amenable to prenatal diagnosis by chorion villus sampling.

2.3 Genetic counselling

Definition

Over the years several definitions have emerged. Some of these tend to be rather wordy and ethereal, while others are so concise as to be incomplete. A reasonable compromise, which includes most of the relevant points, reads as follows: 'Genetic counselling is a communication process by which patients and/or their relatives who may be at risk of an inherited disorder are informed of the consequences of the disorder [and] the likelihood of developing and transmitting it and of ways in which this may be prevented' (Harper 1988).

It is apparent that the essence of genetic counselling is the provision of information. It cannot be overstressed that genetic counselling should never be directive. The counsellor's role is to offer information about risks and options, rather than to direct patients along a particular path of action. On occasions it can be very difficult for the counsellor not to convey his or her own sentiments. Indeed, some patients seem to welcome this, and press for guidance. It must be remembered, however, that it is the patient, not the counsellor, who has to live with the consequences of the decision; so that great care should be taken before personal opinions are expressed.

Requirements for genetic counselling

Most doctors and paramedical staff probably feel that they are perfectly capable of conveying relevant information to patients. Consequently it might be argued that genetic counselling requires little special expertise, and that it can be readily carried out by anyone reasonably familiar with the relevant facts. At the risk of offending colleagues the author would like to outline some of the basic requirements for conveying sensitive information, and to highlight some of the pitfalls. The interested reader is invited to pursue this subject in depth elsewhere (Emery and Pullen 1984).

The setting: Any discussion of important issues—and what could be more important than the health of one's future children—should be undertaken in a congenial, relaxed atomosphere in pleasant surroundings. Ideally a quiet room far removed from a noisy ward or busy out-patient clinic should be used, and if at all possible it is desirable to convey an impression, however false, of unlimited time.

Privacy and confidentiality: If an anxious patient is going to enter into a frank discussion it is vital that he or she can speak freely, in the knowledge that confidentiality will be respected. Someone with a family history of a stigmatizing disorder such as Huntington's chorea may well wish to conceal this information from hospital and general practitioner records for fear of jeopardizing future employment prospects. The presence of interested medical students or nurses, however well intended, may have a profoundly inhibiting effect on open discussion.

Knowledge and information: It is essential that the counsellor should be fully conversant with all the medical and genetic facts relating to the patient's condition, and that the diagnosis has been correctly established. As a result of enthusiastic lay support groups and medical columns in the popular press, members of the general public are often better informed about recent discoveries than their medical advisers! Given that over four thousand single-gene disorders have been described it is clearly impossible for a busy general practitioner or obstetrician to keep abreast of developments. Time spent in preparation for a genetics clinic often exceeds the actual duration of the counselling session.

Communication skills and empathy: It can be very difficult to avoid 'talking down' to patients, yet it is vitally important to avoid technical jargon and a patronizing approach if meaningful dialogue is to take place. A counsellor learns much more about a patient by listening than by talking. Insight into human aspirations and anxieties should ensure that unvoiced questions and concerns can be raised in a gentle and sensitive manner.

Provision of information: Information about risks must be delivered and reiterated in a clearly comprehensible fashion. Dreadful mistakes have resulted when patients have misinterpreted data about risks, assuming for example that a risk of 1 in 2 implies that only alternate children will be affected. At many genetics clinics it is now customary for the patient to be sent a letter summarizing relevant points relating to risks and available options.

Long-term contact: A follow-up visit from a genetics nurse or field-worker is often helpful for reinforcement, and to answer queries which have occurred in the interim. Often such individuals will be seen as non-threatening intermediaries between the 'awe-inspiring' consultant and the 'humble' patient. Genetic registers provide a very useful means for maintaining informal contact between patients and the genetics centre, so that each may approach the other if, for example, an unexpected pregnancy occurs or new information about a disease is forthcoming.

2.4 Genetic counselling and chorion villus sampling

Pre-pregnancy counselling

In an ideal situation it is highly desirable for parents who are believed to be at increased risk of having an abnormal child to be fully assessed and investigated well in advance of pregnancy. This allows time for cytogenetic and/or molecular studies to be undertaken to ensure that prenatal diagnosis is in fact indicated and wanted, and that chorion villus sampling will provide the required information.

For example many women at their first visit to the antenatal clinic mention a family history of Down syndrome, which may prompt a request for prenatal diagnosis. To assess this situation accurately it is usually necessary to establish the family history, and to obtain details of cytogenetic studies in the affected relative or in the mother herself. Often there will be no increased risk; but several weeks may elapse before this can be determined. Similarly for many single-gene disorders, such as cystic fibrosis or Duchenne muscular dystrophy, it is vital to undertake complex molecular studies as outlined in Chapter 3 before it is possible to determine whether prenatal diagnosis can be offered.

Thus, if at all possible, prenatal diagnostic counselling should take place before conception. This also permits time for the various prenatal diagnostic technical options to be considered along with their risks and limitations, as described in the next few paragraphs.

Counselling for chorion villus sampling

As far as the test itself is concerned, the counsellor should bear these key points in mind when discussing the procedure:

Risks: No-one wishes to over-emphasize these, but, if only for the sake of the beleaguered medical defence organizations, it is obviously important that reference is made to the possible risks and complications of any investigation. Thus for chorion villus sampling this must include at the very least discussion of the possibility of miscarriage. In addition it would also seem prudent to alert the patient to the possibility that the test may be unsuccessful for technical or laboratory reasons, so that a repeat investigation or subsequent amniocentesis or cordocentesis may be indicated. Perhaps it is not essential to allude to unusual problems, such as maternal cell contamination, mosaicism and pseudomosaicism, or the unexpected discovery of a sex-chromosome anomaly—although if something of this nature does occur many patients are angry that they were not forewarned of the possibility.

Limitations: Understandably a woman may be under the impression that a normal test result guarantees a normal baby. Thus it should be pointed out that only specific conditions are being investigated, and that numerous congenital abnormalities, particularly neural tube defects, will escape detection. When chorion villus sampling is being undertaken for molecular or biochemical studies, if at all possible it should be clearly established in advance whether chromosomes will also be examined. When prenatal diagnosis for a single-gene disorder is being based on analysis of linked DNA markers, as discussed later in this chapter, it is vital that the patient appreciates the small risk of error through recombination between the disease gene and adjacent marker *loci.*

Maintenance of contact: It is always desirable to establish before chorion villus sampling is undertaken how, when, and by whom the results will be conveyed, particularly since time is often critical if termination of pregnancy is then requested. Providing a telephone number for contact if the patient has not been approached within a specified time is one way of offering some peace of mind during what is likely to be a very anxious wait. It is also kind to discuss how the patient should respond in the unlikely event that she experiences abdominal pain or vaginal bleeding following the test. If the pregnancy is terminated, some form of follow-up counselling and support, with home visits from a nurse or midwife, is usually greatly appreciated, although experience with amniocentesis suggests that this is not always forthcoming (Lloyd and Laurence 1985). Finally, if the pregnancy proceeds arrangements should be made to ensure that the baby is examined shortly after birth by a member of the paediatric staff who is aware of the family history.

2.5 Disorders which can be diagnosed by chorion villus sampling

Of the three classes of inherited disorders outlined earlier in this chapter, only two, chromosomal and single-gene disorders, can be diagnosed using chorion villi. Ongoing research strongly suggests that some conditions in the remaining multifactorial category may also soon be amenable to prediction by analysis of 'susceptibility' genes; but it is likely to be several years before this approach has a clearly demonstrated clinical application.

Chromosome abnormalities

As will be extensively discussed in Chapter 7, chromosome studies can be undertaken on chorion villi using several techniques. In general terms gross structural abnormalities, such as triploidy, trisomy, and monosomy, can be readily identified. If good-quality preparations are obtained, smaller balanced and unbalanced rearrangements can usually be detected. Very subtle chromosome abnormalities, involving loss or gain of tiny amounts of chromosome material, can be very difficult to detect by chorion villus sampling, so that the prenatal diagnosis of microdeletion syndromes such as the Prader–Willi syndrome (15q11.2), Langer–Giedion syndrome (8q23.3), and Miller–Dieker syndrome (17p13.3) should be attempted with great caution, and only after lengthy discussion with cytogenetic colleagues.

Similarly, chromosome studies of chorion villi are of dubious benefit for the prenatal diagnosis of the Fragile X syndrome, which is best diagnosed during pregnancy by cytogenetic studies on lymphocytes obtained by fetal blood-sampling. Ultimately, the molecular approach may permit reliable first-trimester prenatal diagnosis of this very important cause of non-specific mental retardation.

The indications for prenatal diagnostic chromosome studies using chorion villi are listed in Table 2.1. Advanced maternal age is by far the most common indication. Maternal age-specific rates for trisomy 21 are well-established for amniocentesis and live births. Figures for chorion villus sampling are now becoming available, revealing, as indicated in Table 2.2, a spontaneous loss of approximately 20 per cent of Down syndrome conceptions between the time of chorion villus sampling and amniocentesis (Hook *et al.* 1988; Schreinemachers *et al.* 1982). It should be noted that the values in Table 2.2 should be increased by approximately 50 per cent to include all chromosome anomalies.

In the past it has been suggested that advanced paternal age is independently associated with increased risk of chromosome abnormality, but accumulating evidence suggests that any effect is minimal or non-existent (Ferguson-Smith and Yates 1984).

Parents who have had one child with an apparently *de novo* chromosome abnormality run an increased risk for having another baby with some form

Table 2.1. Indications for chromosome studies using chorion villus sampling

Advanced maternal age (> 35 years)
Previous child with chromosome abnormality
Abnormal parental karyotype, such as
inversion
reciprocal translocation
Robertsonian translocation
Fetal sexing for X-linked disorders
Maternal anxiety

Table 2.2. Maternal age-specific rates for trisomy 21 (per 1000 pregnancies)

Maternal age in years	Chorion villus sampling*	Amniocentesis†	Liveborn infants†
35	4.2	4.0	2.6
36	5.7	5.2	3.3
37	7.5	6.7	4.4
38	10.0	8.7	5.7
39	13.4	11.2	7.3
40	17.9	14.5	9.4
41	23.8	18.9	12.3
42	31.7	24.4	15.6
43	42.3	32.3	20.0
44	56.4	40.0	26.3
45	75.1	52.6	33.3

*Hook *et al.* (1988).
†Schreinemachers *et al.* (1982).

of chromosome abnormality. If the index case has trisomy 21, then the figure derived for recurrence of any chromosome anomaly is approximately 1.5 per cent with just over half this risk being for a recurrence of trisomy 21 (Stene *et al.* 1984).

If one parent carries a chromosome rearrangement, such as a Robertsonian translocation, in a balanced form, then it is possible that pregnancy will result in an infant with an unbalanced chromosome complement. Generally this risk will be higher if it is the mother who carries the balanced rearrangement. Risk figures for various rearrangements have been derived from a very large European collaborative study (Boué and Gallano 1984), and are indicated in Table 2.3. It should be stressed that these figures may vary depending on the mode of ascertainment, so that recourse to expert opinion is always advisable when counselling in these situations.

The final indication for chorion villus sampling in Table 2.1 is maternal anxiety. To some extent this reflects the author's own bias, rather than a scientifically valid criterion. Women who have worked with handicapped children are often very anxious about having a child with Down syndrome, and it would seem reasonable to acquiesce to their request for prenatal diagnosis. The author is a firm believer in feminine intuition!

Single-gene disorders

Two approaches can be adopted for the prenatal diagnosis of single-gene disorders using chorionic villi. The first utilizes the more traditional

Table 2.3. Observed outcome in balanced chromosome rearrangements

Parental rearrangement	Number of diagnoses	Karyotype in offspring		
		Normal	Balanced	Abnormal
Inversion				
paternal	51	14	35	2 (4%)
maternal	67	32	30	5 (7.5%)
Robertsonian translocation				
13q14q				
paternal	73	27	46	0
maternal	157	69	88	0
14q21q				
paternal	51	20	31	0
maternal	137	48	68	21 (15.3%)
Reciprocal translocation				
paternal	231	97	107	27 (11.7%)
maternal	378	168	166	44 (11.6%)

Adapted from tables 5, 6, and 10 in Boué and Gallano (1984).

method of biochemical assay, whilst the second and much more novel approach utilizes analysis of DNA.

Biochemical studies

It is rapidly becoming apparent that the diagnosis of any inborn error of metabolism which can be achieved using amniocytes can also be made using chorion villi either directly or following culture. To achieve this, careful dissection is critical to minimize risks of maternal cell contamination; and it is absolutely imperative that the enzyme defect in the index case, usually an affected older sibling, has been accurately identified. In the United Kingdom only a few large laboratories have the expertise and experience necessary to undertake these specialist prenatal diagnostic investigations, so that once again careful planning is vital, particularly if the chorion villi have to be transported for long distances. Some of the rapidly increasing number of metabolic disorders which have been diagnosed by chorion villus sampling and biochemical assay are listed in Table 2.4. This list is not intended to be complete, and anyone presented with a request for prenatal diagnosis of an inborn error should not hesitate to seek advice from the local regional genetics centre or diagnostic biochemical laboratory.

Table 2.4. Inborn errors of metabolism diagnosed by chorion villus sampling and biochemical analysis

Lipid-storage disorders
Fabry's disease*
Gaucher's disease
Krabbe's disease
Metachromatic leukodystrophy
Mucolipidosis II
Niemann–Pick disease
Sialidosis
Mucopolysaccharidoses
Hurler's syndrome
Hunter's syndrome*
Sanfilippo's syndrome (types A and B)
Aminoacidopathies
Cystinosis
Homocystinuria
Maple-syrup urine disease
Tyrosinaemia
Carbohydrate disorders
Fucosidosis
Galactosaemia
Glycogen-storage disease (types II and IV)
Mannosidosis
Others
Lesch–Nyhan syndrome*
Menkes's disease*
Zellweger's syndrome

ALL these disorders show autosomal recessive inheritance apart from those indicated with an asterisk (*).

Molecular studies

The last decade has witnessed a revolution in molecular biology. The ability to isolate sequences of DNA (gene 'probes') and use these to flag an abnormal gene segregating in a family has added a new dimension to the prenatal diagnosis of single-gene diseases. This applies particularly to those numerous disorders in which the basic defect at the protein level is either unknown or not expressed in amniocytes and cultured villus cells. It is this ability by bypass the traditional need for a biochemical handle which has generated such excitement, and welcome in a new era in the study of inherited disease.

A list of some of the more common conditions for which prenatal diagnosis can now be attempted using DNA studies, either by direct analysis of the disease mutation or by gene 'tracking', is provided in Table 2.5. As with the conditions listed in Table 2.4, it should be stressed that this list is far from being complete, with new additions being made on an almost weekly basis. Thus once again it is strongly recommended that any obstetrician confronted with a request for prenatal diagnosis of a single-gene disorder should seek up-to-date information from the local regional genetics centre.

The practical aspects of molecular analysis are considered at length in Chapter 3. At this point it is sufficient to stress that these tests are technically demanding and time-consuming, so that it is always desirable for family assessment to be undertaken well in advance of pregnancy. This is necessary to establish if the family is genetically 'informative', that is, whether particular family members are heterozygous for the linked disease-markers, and if so whether linkage phase can be established, thus permitting subsequent prenatal diagnosis. If these studies are delayed until a future pregnancy is confirmed then there is a very real possibility that existing demands on laboratory services, coupled with the time-consuming nature of the hunt for informative probes, may render first-trimester, and perhaps even second-trimester prenatal diagnosis impossible.

Table 2.5. Single-gene disorders for which prenatal diagnosis can be attempted using DNA analysis

Autosomal dominant
Adult polycystic kidney disease
Huntington's chorea
Myotonic dystrophy
Osteogenesis imperfecta
Retinoblastoma
Autosomal recessive
α1-antitrypsin deficiency
α and β thalassaemia
Congenital adrenal hyperplasia
Cystic fibrosis
Phenylketonuria
Sickle-cell anaemia
Sex-linked recessive
Becker and Duchenne muscular dystrophy
Chronic granulomatous disease
Haemophilia A and B
Norrie's disease
Retinitis pigmentosa (only the X-linked form)

2.6 Risk-calculation and chorion villus sampling

Fundamental to the provision of good genetic counselling is the accurate assessment of risks. Often this will be relatively straightforward, as in uncomplicated autosomal dominant or recessive inheritance. However, on occasions factors such as delayed age of onset, reduced penetrance, and problems of carrier-detection may mean that quite complex calculations have to be undertaken. This subject is discussed at length in several excellent texts (Harper 1988; Stevenson and Davison 1976; Murphy and Chase 1975).

Having established that a pregnancy is at a particular risk, the next step is to determine the predictive error of the tests employed in association with chorion villus sampling. If a gene-specific probe is used which, in combination with a particular restriction enzyme, is able to detect the underlying point-mutation or deletion, then predictive error will be essentially negligible. This applies to sickle-cell anaemia, many cases of α thalassaemia, and, at the time of writing, approximately 50 per cent of cases of Duchenne and Becker muscular dystrophy. For a few disorders a short synthetic 'oligonucleotide' probe can be used to provide very accurate prenatal diagnosis by nature of its ability to hybridize only with the normal allele. This technique has been applied very successfully to the prenatal diagnosis of α1-antitrypsin deficiency (Kidd *et al.* 1984).

The condition of α1-antitrypsin deficiency also illustrates another very important principle. If all of the mutant genes for a particular disease in a specific community or ethnic group originated from a single individual, it may be that adjacent portions of DNA always show the same characteristic pattern. This phenomenon of linkage disequilibrium is illustrated by the Z allele of the α1-antitrypsin gene, which always appears to be associated with a unique closely adjacent polymorphic restriction-enzyme cutting site. If this linkage disequilibrium is very strong, then knowledge of it can be utilized to offer both prenatal diagnosis and carrier detection.

Genetic linkage

Unfortunately, for many conditions, such as Huntington's chorea, gene-specific probes are not available. For these disorders recourse has to be made to genetically linked markers identified by 'restriction fragment length polymorphisms' (RFLPs), which are demonstrated using a gene probe and a DNA cutting (restriction) enzyme whose cleavage site adjacent to or within the disease *locus* constitutes the linked marker.

For the uninitiated the concept of genetic linkage can be bewildering. Put simply, if two genes are linked then their *loci* on the same chromosome are so close together that separation by a cross-over (recombination) at meiosis is very improbable. The likelihood that a cross-over will occur between the two *loci* is known as the recombination fraction (θ), and is denoted as a proportion of 1 (for example, 0.05) or as a percentage (for example, 5 per cent). This subject has been discussed very clearly in a recent review (Yates and Connor 1986).

The use of linked markers in prenatal diagnosis introduces a margin of error which will be related to the likelihood of recombination. Examples of how this approach can be utilized and how predictive errors can be calculated are now given.

Autosomal dominant inheritance

Example 1: Phase known. In Fig. 2.1, A and B are allelic marker genes situated at a *locus* closely linked ($\theta = 0.05$) to the disease *locus*. Careful study of the pedigree reveals that in II2 the disease gene must be coupled with marker B, since both the disease and this marker must have been inherited from I2. Therefore knowing that the disease is in coupling with marker B in II2 enables the disease status for III1 to be predicted. If she inherits her mother's A marker then the probability that she will be affected equals θ (0.05), since she will only be affected if recombination has occurred between the disease gene and marker B during maternal meiosis. Similarly if III1 inherits her mother's B marker then the probability that she will be affected equals $1 - \theta$ (0.95), that is, she will be affected unless recombination has occurred between the disease gene and marker B during maternal meiosis. Thus for this prenatal (or preclinical) test there will be a predictive error of 5 per cent.

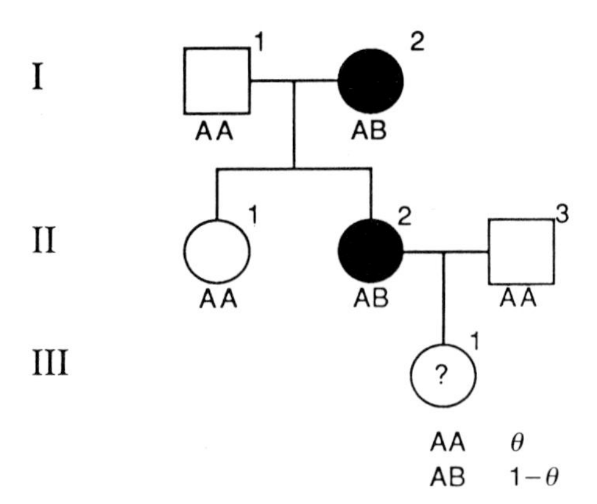

Fig. 2.1. (Example 1) Autosomal dominant inheritance—phase known. The disease gene in II2 must be in coupling with marker B. Thus the risk for any child of II2 can be calculated as $(1 - \theta)$ if marker B is inherited, and θ if marker A is inherited.

Example 2: Phase unknown. In Fig. 2.2 it is readily apparent that the disease in II1 is in coupling with marker B, with both the disease gene and marker B having been inherited on the same chromosome from I2. However, this does not necessarily mean that the disease gene in I2 is on the same chromosome as marker B, as a cross-over could have occurred during meiosis in the gamete which went to form II1. Therefore in determining risks for II2 the calculation has to take into account the possibilities that the disease in I2 is in coupling with either A or B. This is achieved using Bayes's theorem, in which prior information and observed (conditional) information are combined to derive posterior or relative probabilities for each possibility. Bayes's theorem and its application in risk-calculation are discussed at length elsewhere (Emery 1986).

	Disease in I2 in coupling with A	Disease in I2 in coupling with B
Probability		
Prior	$\frac{1}{2}$	$\frac{1}{2}$
Conditional		
II1 is affected and has inherited B	θ	$1-\theta$
Joint	$\frac{\theta}{2}$	$\frac{1-\theta}{2}$

Thus the posterior probability that the disease in I2 is in coupling with marker A equals

$$\frac{\frac{\theta}{2}}{\frac{\theta}{2}+\frac{1-\theta}{2}}$$

which equals θ.

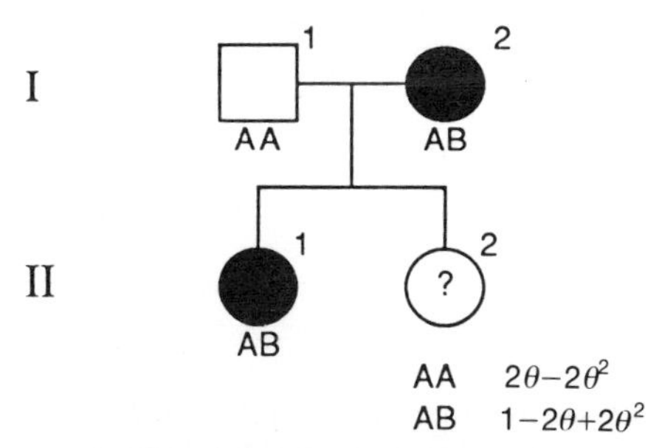

Fig. 2.2. (Example 2) Autosomal dominant inheritance. The linkage phase for I2 is not known, but information about its likelihood can be obtained from II1 as indicated in the text. This then enables risks for II2 to be calculated.

Similarly the posterior probability that the disease in I2 is in coupling with marker B equals $1 - \theta$.

For individual II2 in Fig. 2.2, the calculation proceeds as follows. The probability that she has inherited the disease gene if she has inherited marker A from I2 equals the sum of

(1) the probability that the disease in I2 is in coupling with marker A (θ) multiplied by $1 - \theta$, i.e. $\theta(1 - \theta)$, plus

(2) the probability that the disease in I2 is in coupling with marker B ($1 - \theta$) multiplied by θ, i.e. $\theta(1 - \theta)$.

This summates to $2\theta - 2\theta^2$, which equals 0.095 if $\theta = 0.05$.

The probability that II2 has inherited the disease gene if she has inherited marker B from I2 equals the sum of

(1) the probability that the disease in I2 is in coupling with marker B ($1 - \theta$) multiplied by $(1 - \theta)$, i.e. $(1 - \theta)^2$, plus

(2) the probability that the disease in I2 is in coupling with marker A (θ) multiplied by θ, i.e. θ^2.

This summates to $1 - 2\theta + 2\theta^2$, which equals 0.905 if $\theta = 0.05$.

Therefore in this example, where the linkage phase in I2 is not known with certainty, there is a predictive error of approximately 10 per cent, almost twice that which was calculated for Example 1.

Example 3: Prenatal exclusion diagnosis. In Fig. 2.3 II1 has not yet reached the age of onset of a late-onset autosomal dominant disorder such as Huntington's disease. As her affected father (I1) is homozygous for the closely linked disease-marker, there is no way of predicting disease status

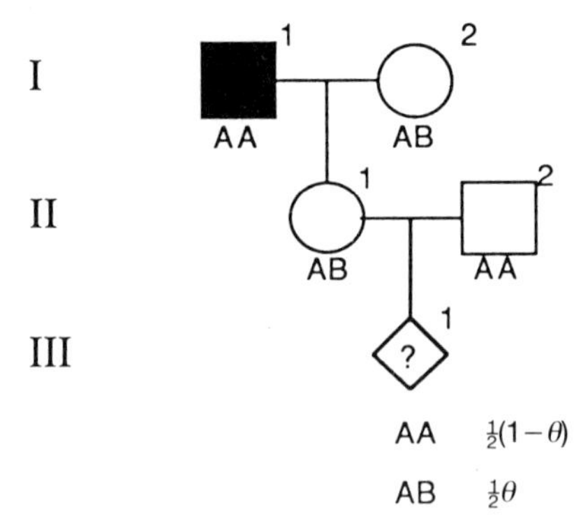

Fig. 2.3. (Example 3) Autosomal dominant inheritance—prenatal exclusion. There is a 50 per cent probability that the A marker gene in II1 is in coupling with the disease gene, so that any fetus inheriting this A gene will be at a risk of $\frac{1}{2} \times (1 - \theta)$ for inheriting the disease gene.

for II1. However, it would be possible to monitor any pregnancy conceived by II1 and II2 to see if the fetus (III1) has inherited marker A from its affected grandfather or marker B from its unaffected grandmother. Marker A would convey a risk of $\frac{1}{2}(1-\theta)$ to the fetus, which will be close to 50 per cent, whereas marker B will convey a risk of $\frac{1}{2}\theta$, that is, a very low risk. In situations such as this some parents may elect to terminate the 'high risk' pregnancy, thereby exercising their option to 'exclude' the diagnosis.

Autosomal recessive inheritance

Example 4: Information from one child. In Fig. 2.4 the first child has an autosomal recessive disorder, and prenatal diagnosis is available using a closely linked marker with alleles A and B. The calculation of the risk that the fetus II2 will be affected given different genotypes AA, AB, and BB has to take into account the posterior probabilities for the carrier haplotypes in the parents, based upon information provided by II1, who is affected and has an AA marker genotype. These posterior probabilities are calculated as shown.

Phase of disease gene in father (F) and mother (M).

	F(A) M(A)	F(A) M(B)	F(B) M(A)	F(B) M(B)
Probability				
Prior	$\frac{1}{4}$	$\frac{1}{4}$	$\frac{1}{4}$	$\frac{1}{4}$
Conditional				
II1 has AA	$(1-\theta)^2$	$(1-\theta)\theta$	$\theta(1-\theta)$	θ^2

From this table it is apparent that the posterior probabilities for each of the 4 different parental disease haplotypes are:

Father A: Mother A—$(1-\theta)^2$
Father A: Mother B—$(1-\theta)\theta$
Father B: Mother A—$\theta(1-\theta)$
Father B: Mother B—θ^2

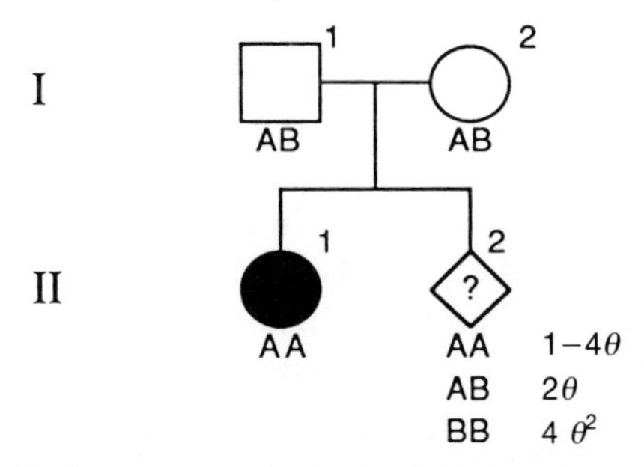

Fig. 2.4. (Example 4) Autosomal recessive inheritance. Approximate values for the probability that the fetus will be affected given different marker genotypes are indicated. See text for derivation of these values.

(1) Fetus (II2) has an AA genotype. The overall probability that the fetus will be affected given an AA genotype will be the sum of

(a) the probability if the disease gene is in coupling with A in both parents, that is

$$(1-\theta)^2 \times (1-\theta)^2, \text{ plus}$$

(b) the probability if the disease gene is in coupling with A in the father and B in the mother, that is

$$(1-\theta)\theta \times (1-\theta)\ \theta, \text{ plus}$$

(c) the probability if the disease gene is in coupling with B in the father and A in the mother, that is

$$\theta(1-\theta) \times \theta(1-\theta), \text{ plus}$$

(d) the probability if the disease gene is in coupling with B in both parents, that is

$$\theta^2 \times \theta^2.$$

This summates to $(1-\theta)^4 + 2\theta^2(1-\theta)^2 + \theta^4$, which equals $1 - 4\theta + 8\theta^2 - 8\theta^3 + 4\theta^4$.

(2) Fetus (II2) has a BB genotype. The overall probability that the fetus will be affected given a BB genotype will be the sum of

(a) the probability if the disease gene is in coupling with A in both parents, that is

$$(1-\theta)^2 \times \theta^2, \text{ plus}$$

(b) the probability if the disease gene is in coupling with A in the father and B in the mother, that is

$$(1-\theta)\theta \times \theta(1-\theta), \text{ plus}$$

(c) the probability if the disease gene is in coupling with B in the father and A in the mother, that is

$$\theta(1-\theta) \times (1-\theta)\theta, \text{ plus}$$

(d) the probability if the disease gene is in coupling with B in both parents, that is

$$\theta^2 \times (1-\theta)^2.$$

This summates to $4\theta^2(1-\theta)^2$, which equals $4\theta^2 - 8\theta^3 + 4\theta^4$.

(3) Fetus (II2) has an AB genotype. In this situation the fetus could have inherited A from father and B from mother, or vice versa. If it is assumed that the A has come from the father and the B from the mother, then the overall probability that the fetus will be affected will be the sum of

(a) the probability if the disease gene is in coupling with A in both parents, that is

$$(1-\theta)^2 \times (1-\theta)\theta, \text{ plus}$$

(b) the probability if the disease gene is in coupling with A in father and B in the mother, that is

$$(1-\theta)\theta \times (1-\theta)^2, \text{ plus}$$

(c) the probability if the disease gene is in coupling with B in the father and A in the mother, that is

$$\theta(1-\theta) \times \theta^2, \text{ plus}$$

(d) the probability if the disease gene is in coupling with B in both parents, that is

$$\theta^2 \times \theta(1-\theta).$$

This summates to $2\theta(1-\theta)^3 + 2\theta^3(1-\theta)$, which equals $2\theta - 6\theta^2 + 8\theta^3 - 4\theta^4$.

An identical result will be obtained if the fetus has inherited A from the mother and B from the father. Thus, whichever way the AB genotype has been derived, it conveys a probability of $2\theta - 6\theta^2 + 8\theta^3 - 4\theta^4$ for being affected.

Since θ is likely to be a very small number (for example, less than 0.01 for most of the linked markers used in the prenatal diagnosis of cystic fibrosis) reasonable approximations for the probability that the fetus will be affected are

Genotype AA—$1-4\theta$
AB—2θ
BB —$4\theta^2$

Example 5: Information from two children. If the linkage information is concordant in two children, as in Fig. 2.5, then it can reasonably be assumed that the linkage phase in the parents is known. Thus in Fig. 2.5 the disease gene in both parents can be taken as being in coupling with marker gene A.

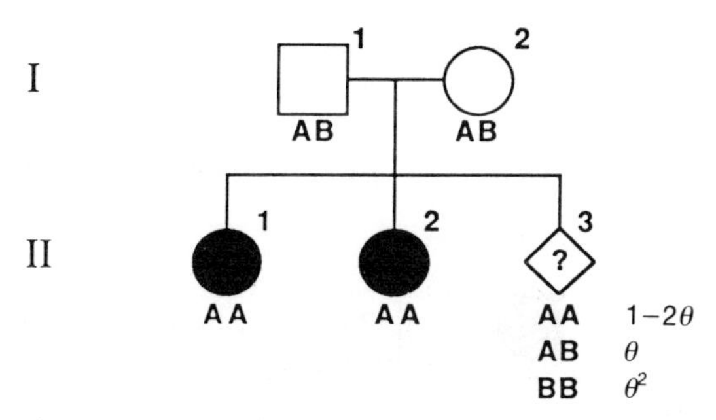

Fig. 2.5. (Example 5) Autosomal recessive inheritance. In this situation concordant linkage information from two affected children allows the linkage phase in the parents to be assumed. Approximate values for the probability that the fetus will be affected given different marker genotypes are indicated.

Risks for the next child are calculated as previously. Thus if the fetus inherits an AA genotype the likelihood that it will be affected equals $(1-\theta)^2$, which equals $1-2\theta+\theta^2$, and approximates to $1-2\theta$. If the fetus inherits an AB genotype the likelihood that it will be affected equals $(1-\theta)$ (for marker A) $\times\ \theta$ (for marker B). This equals $\theta-\theta^2$, and an approximate value of θ can be applied. If the fetus inherits a BB genotype then the likelihood that it will be affected equals θ^2.

Thus the predictive errors for these various marker genotypes will be

AA 2θ for not being affected
AB θ for being affected
BB θ^2 for being affected.

Sex-linked recessive inheritance

Example 6: In Fig. 2.6, the mother (II2) of a boy affected with a sex-linked recessive disorder requests prenatal diagnosis during her next pregnancy. A and B represent marker genes closely linked to the disease *locus*. The calculation of risks for a future son (III2) given different marker genotypes proceeds as follows.

The mother II2 must be a carrier, since she has had two affected brothers and an affected son. Thus the possibility that the affected son (III1) represents a new mutation or has inherited his disease as a consequence of maternal gonadal mosaicism can be ignored. Unfortunately individuals I1, I2, II3, and II4 are no longer alive, so that the linkage phase in the mother is not known with certainty. Thus the situation is very similar to that discussed in Example 2, and risks for III2 are calculated in an identical fashion.

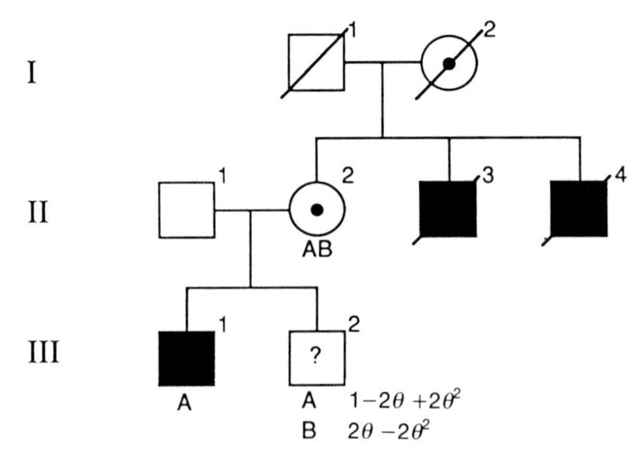

Fig. 2.6. (Example 6) Sex-linked inheritance, phase not known. This situation is essentially the same as that outlined in Example 2. Values for the probability that a male fetus will be affected given each marker genotype are indicated.

The probability that marker gene A is in coupling with the disease in II2 equals 1-θ. Similarly the probability that marker gene B is in coupling with the disease in II2 equals θ. Thus if III2 inherits marker A from his mother, the probability that he will also inherit the disease equals $(1-\theta)^2+\theta^2$, which summates to $1-2\theta+2\theta^2$. If III2 inherits marker B from his mother, the probability that he will also inherit the disease equals $\theta(1-\theta)+(1-\theta)\theta$, which summates to $2\theta-2\theta^2$. The full derivation of these risks is given in Example 2. As in Example 2 there will therefore be a predictive error of approximately 10 per cent if $\theta=0.05$.

Comments on genetic linkage

It will be apparent to the discerning reader that linkage analysis can generate difficult calculations! In fact those outlined in the preceding pages illustrate relatively simple examples. The use of flanking markers, that is, markers one on each side of the disease *locus*, enables greater diagnostic precision, but the calculations become much more complex (Winter 1985). With sex-linked recessive disorders allowance may have to be made for the possibility that the disorder has arisen as the result of a new mutation in an isolated case, so that the results of carrier-detection tests may have to be taken into account (Pembrey *et al.* 1984). Many centres actively involved in these investigations now use computer programmes to solve some of the more difficult calculations.

It should also be stressed that non-paternity will often invalidate linkage analysis, so that a case can be made for confirming family structure using DNA minisatellite genetic 'fingerprinting' (Jeffreys *et al.* 1985). Finally, it is important to remember that the diagnosis in a fetus aborted as a consequence of DNA linkage studies cannot be confirmed if knowledge of the protein defect is missing. This emphasizes the importance of the traditional biochemical approach, and the desirability of using methods which directly identify the disease mutation.

2.7 Conclusion

In this chapter an attempt has been made to outline some of the more important aspects of genetic counselling as they relate to chorion villus sampling. The spectrum of inherited disease has been reviewed, along with the role of genetic counselling both in general terms and with regard to prenatal diagnosis. The range of disorders which can be diagnosed has also been discussed. The complex nature of the many issues involved leaves little doubt that the interests of both mother and fetus are best served by close collaboration between all clinicians and scientists actively involved in the provision of these services.

References

Boué, A. and Gallano, P. (1984). A collaborative study of the segregation of inherited chromosome structural rearrangements in 1356 prenatal diagnoses. *Prenatal Diagnosis*, **4**, 45–67.

Emery, A. E. H. (1986). *Methodology in medical genetics*, 2nd edn. Churchill Livingstone, Edinburgh.

Emery, A. E. H. and Pullen, I. (1984). *Psychological aspects of genetic counselling.* Academic Press, London.

Ferguson-Smith, M. A. and Yates, J. R. W. (1984). Maternal age specific rates for chromosome aberrations and factors influencing them: report of a collaborative European study on 52,965 amniocenteses. *Prenatal Diagnosis*, **4**, 5–44.

Harper, P. S. (1988). *Practical genetic counselling*, 3rd edn. John Wright, Bristol.

Hook, E. B., Cross, P. K., Jackson, L., Pergament, E., and Brambati, B. (1988). Maternal age—specific rates of 47, + 21 and other cytogenetic abnormalities diagnosed in the first trimester of pregnancy in chorionic villus biopsy specimens: comparison with rates expected from observations at amniocentesis. *American Journal of Human Genetics*, **42**, 797–807.

Hsu, L. Y. F. (1986). Prenatal diagnosis of chromosome abnormalities. In *Genetic disorders and the fetus*, 2nd edn, ed. A. Milunsky, Chapter 5. Plenum Press, New York.

Jeffreys, A. J., Wilson, V., and Thein, S. L. (1985). Hypervariable 'minisatellite' regions in human DNA. *Nature*, **314**, 67–73.

Kidd, V. J., Golbus, M. S., Wallace, R. B., Itakura, K., and Woo, S. L. C. (1984). Prenatal diagnosis of α1-antitrypsin deficiency by direct analysis of the mutation site in the gene. *New England Journal of Medicine*, **310**, 639–42.

King's Fund Forum (1987). *Screening for fetal and genetic abnormality* (consensus statement). King Edward's Hospital Fund, London.

Lloyd, J. and Laurence, K. M. (1985). Sequelae and support after termination of pregnancy for fetal malformation. *British Medical Journal*, **290**, 907–9.

McKusick, V. A. (1988). *Mendelian inheritance in man*, 8th edn. Johns Hopkins University Press, Baltimore.

Murphy, E. A. and Chase, G. A. (1975). *Principles of genetic counselling.* Year Book Medical, Chicago.

Pembrey, M. E., Davies, K. E., Winter, R. M., Elles, R. G., Williamson, R., Fazzone, T. A., *et al.* (1984). Clinical use of DNA markers linked to the gene for Duchenne muscular dystrophy. *Archives of Disease in Childhood*, **59**, 208–16.

Schreinemachers, D. M., Cross, P. K., and Hook, E. B. (1982). Rates of trisomies 21, 18, 13 and other chromosome abnormalities in about 20,000 prenatal studies compared with estimated rates in livebirths. *Human Genetics*, **61**, 318–24.

Stene, J., Stene, E., and Mikkelsen, M. (1984). Risk for chromosome abnormality at amniocentesis following a child with a non-inherited chromosome aberration. *Prenatal Diagnosis*, **4**, 81–95.

Stevenson, A. C. and Davison, B. C. C. (1976). *Genetic counselling*, 2nd edn. Heinemann Medical, London.
Winter, R. M. (1985). The estimation of recurrence risks in monogenic disorder using flanking marker loci. *Journal of Medical Genetics*, **22**, 12–15.
Yates, J. R. W. and Connor, J. M. (1986). Genetic linkage. *British Journal of Hospital Medicine*, **36**, 133–6.

3 Gene analysis in humans

Roger Quaife

3.1 Introduction

Until the advent of the molecular techniques now being applied, there had been little or no progress in determining the cause of many important genetic disorders. Carriers of such gene disorders could not be reliably identified, nor could the diseases be prevented or cured. Now, however, the molecular defect within the gene has been located for some, and it is likely that others will follow. Perhaps of more import to the clinician is that it is already possible to identify carriers, and to carry out prenatal diagnosis in many genetic disorders while their biochemistry and molecular pathology remain unknown. The principles and methods used are discussed below, followed by some diagnostic examples.

3.2 Basic principles

DNA and gene structure

The reader is reminded that DNA is a right-handed double helix. Each 'strand' of the helix is a sugar–phosphate molecule running in opposite directions, and is joined to the other by hydrogen bonds between the only variable part of the molecule, which is the sequence of the nucleotide bases. These are: Adenine (A) linked to Thymine (T), and Guanine (G) linked to Cytosine (C). Since A only pairs with T and G with C the two strands of the duplex are complementary. Of particular note is that each GC pair is held together by three hydrogen bonds, whereas only two hold the AT pairs. When double-stranded DNA molecules are heated to high temperatures (near 100°C) their hydrogen bonds break, and the two strands separate—the DNA is said to be denatured. Since GC base-pairs are held together more firmly than AT pairs, higher temperatures are required to separate GC-rich strands of DNA. At intermediate temperatures or when destabilizing agents (for example, formamide) are added, DNA is partially denatured. Even complete denaturation is not irreversible. Cooling will result in renaturation, to re-form a double-stranded DNA molecule.

Genetic information is stored within the DNA in the form of a code. This is a sequence of three bases (a codon) which determines the structure of one amino acid. However, not all of the bases in a coding gene code for an amino acid; there are intervening sequences (introns), which are highly variable in length and sequence, but which in most vertebrates are not 'required' for protein coding. Introns are spaced between the groups of coding bases (exons); these act as a DNA template for the synthesis of messenger RNA (mRNA). Initially the RNA copy contains introns and exons; but subsequent processing removes the intron bases and joins the remaining exons together. At this stage the mRNA has a long string of adenylic acid residues added at one end—the poly-A tail. In addition to exons and introns, genes also have regions known as the TATA and CCAAT boxes, and these are involved in the control of RNA transcription.

Gene libraries, cloning, and probes

There are two basic types of gene library: genomic and complementary DNA (cDNA). A genomic library consists of human DNA cut up into randomly sized pieces, and joined (ligated) to the DNA of a replicating vehicle (a vector), such as a bacteriophage. Ideally all the DNA is represented in the library, because the fragments are randomly sized, and should overlap. The cutting or cleavage is done by restriction enzymes. These are bacterial enzymes which can cut double-stranded DNA; the position at which the DNA is cut is determined by the enzyme recognizing its own specific base-sequence. Each vector containing a fragment of DNA gives rise to a clone when it reproduces, each member containing exactly the same genetic information. The number of different clones needed to represent a human as a genomic DNA library depends on the size of the fragments initially cloned. If for example the fragments were of 5000 base-pairs then 600 000 clones would be needed.

A cDNA library represents the mRNA population within the tissue of origin, and from this cDNA's of interest can be selected. Classically the method depends on first extracting mRNA, and, by the use of an enzyme called reverse transcriptase, making a complementary DNA copy of the original mRNA. Further enzyme steps allow a double-stranded DNA copy of the original messenger RNA to be created, and this can then be cloned. Each type of library has advantages. A genomic library should have a copy of every gene present in the original organism; whereas a cDNA library can only represent genes which are transcribed in the tissue of origin of the organism. However, isolating a cDNA clone (representing an individual messenger RNA) can be a relatively easy way to identify the *complete* gene.

Probes are clones that have been radioactively tagged, and can therefore be used to search out DNA sequences complementary to themselves. If both the probe and the DNA to be screened are each first made single-stranded, then under the correct condition they will join (hybridize) to create a double-stranded molecule. The prior radioactive tagging of the probe allows it to pin-point the DNA sequence of interest.

Restriction endonucleases and Southern blotting

The pin-pointing of a particular DNA sequence of bases from an individual's entire DNA complement (genome) requires techniques of the utmost sensitivity. The human genome consists of $3.3–3.5 \times 10^9$ base-pairs, so that a small gene of say 3000 base-pairs makes up only about a millionth of the total DNA present.

The first step is the extraction of DNA. This requires lysis of the cell and the nucleus, so that the DNA is released. A series of purification steps follows, and then the reduction of the enormously long DNA molecules into fragments of defined length. This can be achieved by use of the previously mentioned restriction enzymes, which cut DNA. If the reaction conditions are correct these enzymes will *invariably* cut at their specific points. Hundreds of thousands of different-sized fragments may be produced, but can be separated according to size by agarose gel electrophoresis. DNA is a negatively charged molecule, and under an applied voltage will move to the positive pole. Larger fragments remain near the origin, and the smaller the fragment the further they move. The result is a continuous smear of different-sized DNA fragments on the agarose gel. These may be transferred and fixed permanently by blotting the gel with a nylon or nitrocellulose membrane—a technique invented by Professor Southern, and hence referred to as 'Southern blotting'. Once such membranes have been incubated with a radioactive probe under hybridization conditions the membrane can be exposed to radiation-sensitive film. The photographic emulsion is exposed to radiation only where the probe has bound to a sequence or sequences with which it is partially or completely complementary. The film will have a dark band at these points when developed.

RFLPs, deletions, and duplications

The human genome exhibits a large amount of variation between individuals within a given population. Such variation occurs both within and outside the base-sequences which code for polypeptides. Although the base differences within the coding regions between individuals are relatively rare, it has been calculated that as many as 1 in 100 non-coding nucleotide sites may vary. Some of these variations may alter the site at which a restriction enzyme cuts. The result is that variations are seen in the length of fragments recognized by particular probes. Such variations are known as Restriction Fragment Length Polymorphisms (RFLPs). Most RFLPs probably represent benign changes in the DNA, although some are undoubtedly associated with pathological conditions.

RFLPs can aid carrier and prenatal diagnosis of genetic disease in three ways: firstly, the change in the DNA which gives rise to a disease may also coincidentally alter the site for a restriction enzyme. This is the case in sickle-cell anaemia, which is discussed further below. Secondly, an RFLP may be found in a gene which is abnormal and causes a disease. In this case the RFLP is so close to the disease mutation that it is unlikely to be separated from it at meiosis by crossing-over. An example can be found in the haemophilia B gene. Finally, an RFLP may be recognized by a probe which is actually outside the gene in question, but which is positioned so that both gene and RFLP are inherited together, they are *linked*, and estimates can be made as to how close this linkage is. Both adult polycystic kidney disease and cystic fibrosis are examples.

Major deletions of DNA can readily be identified using restriction enzymes. If a gene is deleted, then the restriction fragment produced will be altered in size. β° thalassaemia is a common example, and has a 600-base-pair deletion at one end of the β globin gene. In this case the restriction sites for the enzyme βgl II are 600 base-pairs closer together than is the case in a non-deleted gene. Similar results can be seen in Duchenne and Becker muscular dystrophies (DMD and BMD), where these diseases can be seen to be caused by gene deletions in at least 67 per cent of cases.

In both DMD and BMD gene duplications have recently been observed. These duplications were found to be within the gene, and probably caused the disease by disrupting the synthesis of mRNA from the gene. Again, like deletions, gene duplications can be detected as a restriction fragment of altered size.

3.3 Some diagnostic examples

The examples given below are designed to demonstrate the principles of single-gene analysis described above. It seems likely that for the near future the most common form of analysis will be by linkage of an anonymous probe sequence (recognizing an RFLP) to the disease of interest.

For prenatal diagnosis the preferred source of fetal DNA is chorionic villi. We (Quaife and Liu 1987) have found by regression analysis that a linear relationship exists between the amount of villus and the subsequent quantity of DNA extracted (for each milligram of villi we would expect a yield of 1.7 μg of DNA). A prenatal diagnosis will require between 5 and 10 μg of fetal DNA, with say 5μg as a back-up. A villus sample should thus be around 10mg or more wet weight. An analysis using standard techniques will take about five days from the time the tissue sample arrives at the laboratory, providing that, if necessary, family members have been previously analysed. The special advantage of using chorionic villi is that fetal DNA (and of course chromosomes) can be obtained and analysed within the first trimester of pregnancy. DNA and chromosome analysis can be performed using cells from the amniotic fluid; but these would need to be cultured, which means that the pregnancy would need to continue to about 19 or 20 weeks before a result would be available. This delay can prove extremely distressing both for the parents and for hospital staff should a termination be required. A recently introduced technique (described below) called 'gene amplification' (or 'the polymerase chain reaction') will reduce the analysis time, and, if used in conjunction with chorionic villus sampling, will have great attractions for parents, clinicians and scientists. Unfortunately, it does require that the base-sequence of the gene is known, and so it will be some time yet before it is used widely in prenatal diagnosis.

Sickle-cell anaemia and gene amplification

In 1978 Kan and Dozy (1978) reported the first prenatal diagnosis of sickle-cell anaemia using a restriction enzyme. The enzyme used then has since been replaced by Mst II, but the principle is the same. The base-sequence CCTGAGG within the region coding for the amino acids Proline–Glutamine–Glutamine (*CCT—GAG—G*AG) is recognized by the enzyme Mst II. However, in sickle-cell disease the codon for one of the glutamines has mutated such that it codes for valine. Not only does this cause the disease, but it also results in the loss of the Mst II cutting sequence. Normal and abnormal genes therefore produce different-sized restriction fragments. In fact a normal gene gives rise to a 200 and a 1150bp fragment. A sickle gene can generate only a 1350bp fragment.

The Polymerase Chain Reaction (PCR) (Saiki *et al.* 1985) method, as an alternative analysis technique, continues to use Mst II to detect differences in the DNA sequence between sickle and normal genes. However, it does not rely on autoradiography or radioactive labelling, since a probe is not used to detect the restriction fragment difference. Rather the gene-sequence is repeatedly synthesized using the heat-stable enzyme Taq polymerase. The reaction is exponential, so that after 1–2 hours a gene-sequence can be reproduced (amplified) by a factor of more than 10 million, with a very high specificity.

PCR amplification uses two synthetically produced sequences of DNA, called oligonucleotide primers (usually about 20 bases long), and each is able to hybridize with opposite strands of the DNA duplex. They are made such that they flank the DNA region to be amplified. The first step involves heating the DNA, which denatures it to the single-stranded state. Some cooling is then allowed, such that the primers can anneal *exactly* to the sequences of bases on the strand for which they were designed. Taq polymerase can then add bases beyond the point at which the primers have annealed. The result is a new strand of that segment of DNA is synthesized. The process can be repeated, and, since the newly synthesized DNA is also complementary to primers, the reaction proceeds exponentially (Fig. 3.1). Once the reaction has been completed, the DNA products can be analysed by gel electrophoresis, (Fig. 3.2). In the example shown here, to detect the sickle mutation a 294bp segment of the β globin gene was amplified. An Mst II digest will produce two fragments of 191bp and 103bp in a normal gene, but only a 294bp fragment in a sickle gene (Fig. 3.3, track 2). Carriers of the disease will have one normal gene and one sickle gene, and so three different-sized fragments, of 294, 191, and 103bp, are seen in these individuals (Fig. 3.3, tracks 4 and 6).

The simplicity and speed of this technique will make it an attractive diagnostic tool as more genes are sequenced. Automation can already eliminate the labour of the thermal cycling, so that potentially this technique would allow population screening for carriers of single-gene defects such as cystic fibrosis.

An intragenic RFLP

Genetic crossing-over (recombination) between a restriction fragment site and a gene mutation is unlikely if the RFLP site is within the gene. An additional advantage is that it is not necessary to identify the actual gene mutation in order to be able to give helpful genetic advice. This is important, since in β thalassaemia, for example, over forty different mutations are recognized as being responsible for the disease.

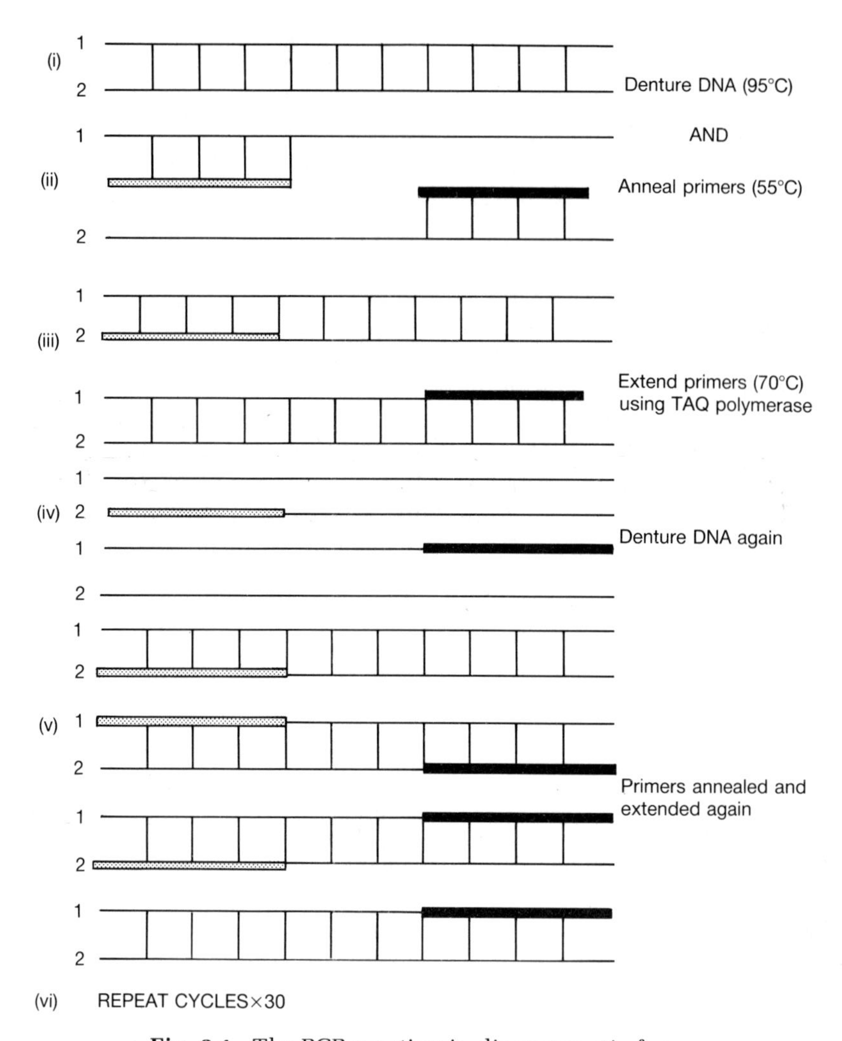

Fig. 3.1. The PCR reaction in diagrammatic form.

Haemophilia B (Christmas disease) is an X-linked recessive disorder caused by a reduction in the coagulation factor IX. cDNA clones were obtained in 1982 (Choo *et al.* 1982), and these clones were used as gene-specific probes to identify the complete gene. Giannelli and colleagues (1984) then developed a subgenomic probe 2.5 thousand base-pairs in length (2.5Kb), containing exon d (114bp long) and 2.4Kb of intron sequence. This probe is able to identify restriction fragments using the enzyme Taq 1 (Fig. 3.4).

How then can this RFLP be used in detecting carriers and in prenatal

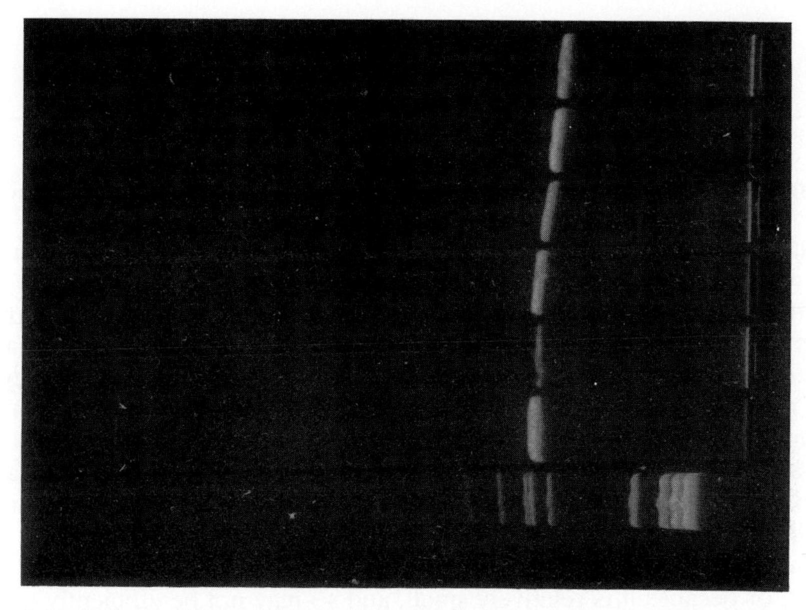

Fig. 3.2. PCR reaction products. Each track shows a single 294bp fragment. The first track is a phi ×174 size marker.

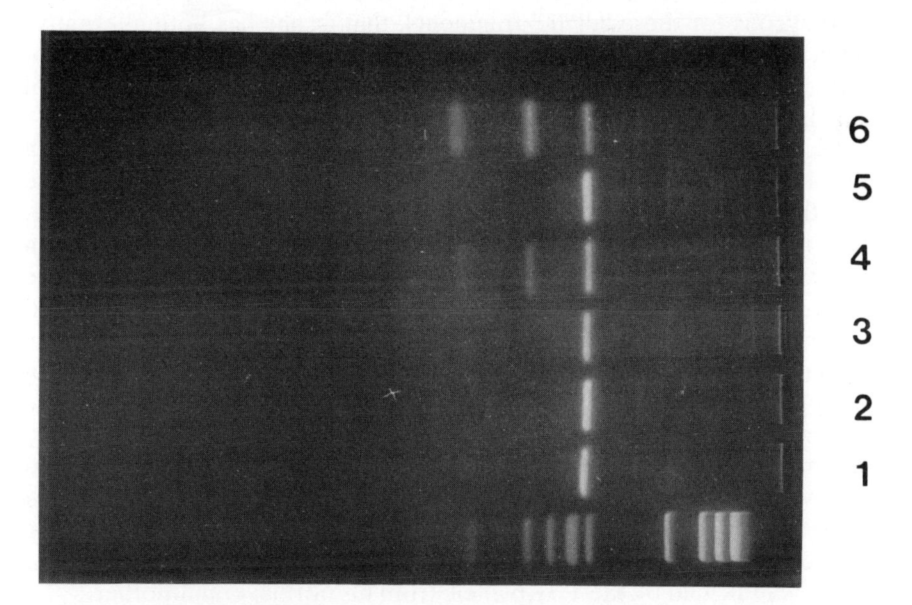

Fig. 3.3. PCR reaction products after an MstII digestion.

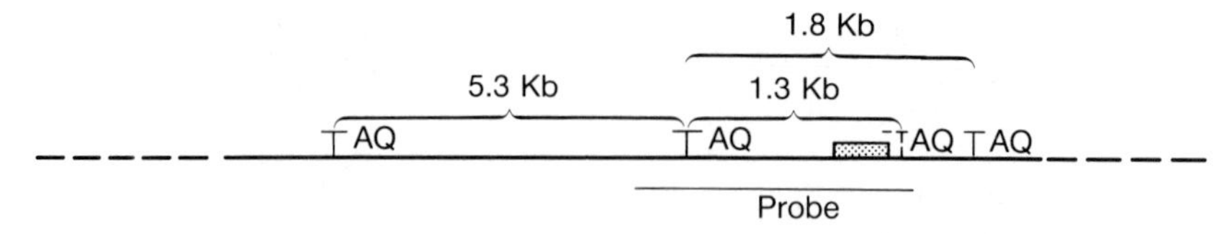

Fig. 3.4. Diagram of the haemophilia B gene, showing Taq 1 restriction sites.

diagnosis? The first prerequisite is that DNA from an affected individual is available. This is because the RFLP is used as a 'marker' to identify the particular X chromosome carrying the abnormal gene. It can be seen from Fig 3.4 that all individuals must have a 5.3Kb fragment, since the two Taq 1 sites producing it do not vary. However, different-sized fragments can be seen around the polymorphic site illustrated (denoted as a broken T in the diagram). Males, because they have a single X chromosome, can have a 1.8 or a 1.3Kb fragment, depending on whether or not the polymorphic Taq 1 site is absent or present. (If the site is present then theoretically a 1.3 and a 0.5Kb fragment should be seen. In practice the 0.5Kb fragment is not observed, either as a result of very little of the probe sequence's hybridizing to it, or because it is relatively small, and so may not be efficiently transferred from the gel to the membrane in Southern blotting.) Females have two copies of the X chromosome, and so fragments of both sizes can potentially be present, again depending on whether the polymorphic site is present or absent on either chromosome. If a female has both, then she is heterozygous for these 'allelic' fragments, that is, she has both alternative fragments. Of course females may have the same 'allele' on both X chromosomes, and so be homozygous. (Males are actually *hemizygous* because of their single X.)

The small pedigree and autoradiograph shown in Fig. 3.5 may be a helpful illustration. The female II_1 must be a carrier for the disease, since her father was affected. He must have given her his only X chromosome, which must also have carried the mutant gene. She is thus established as a carrier. In addition the autoradiograph above the pedigree shows her to be heterozygous—she has fragments of both sizes. Now supposing she wanted to know if her daughter III_1 was also a carrier, and if her fetus III_2 was affected. In fact one can deduce that her daughter is a carrier, since she is a 1.8Kb homozygote, having inherited the defective 1.8Kb X chromosome from her affected grandfather via her mother.

Coincidentally, she has inherited an X chromosome from her normal father which also has a 1.8Kb restriction fragment. This serves to illustrate the point that the RFLP is only a marker, and is not the actual disease mutation. The fetus (III_2) is, however, normal, since he has inherited the normal X indicated by the 1.3Kb allele from his normal grandmother.

When the daughter (III_1) needs a prenatal diagnosis for her fetus, then

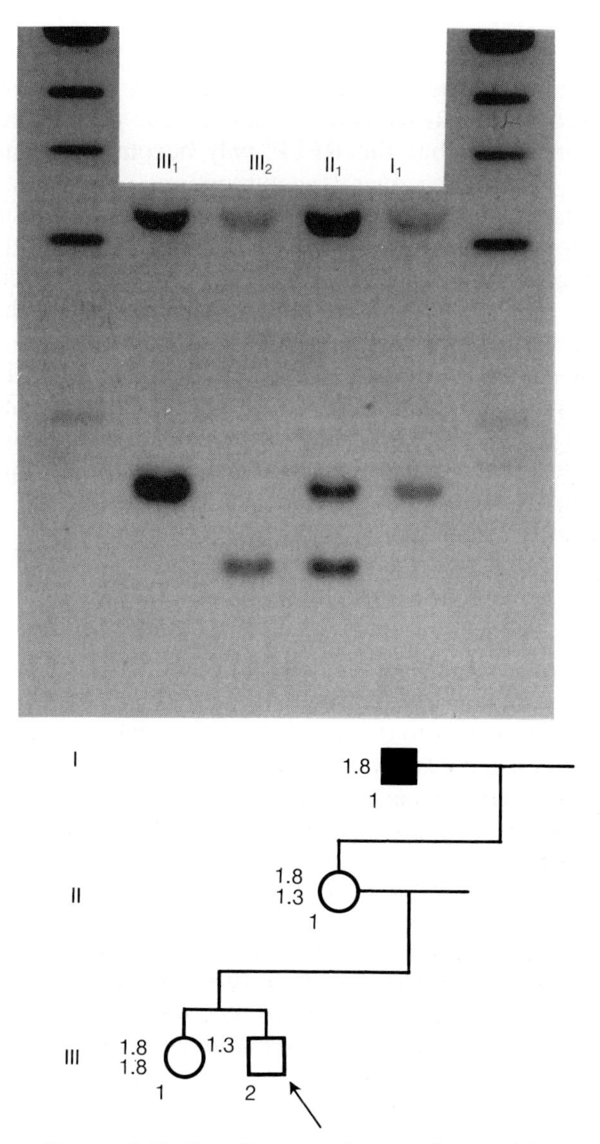

Fig. 3.5. Haemophilia B pedigree and autoradiographic patterns.

the Taq 1 polymorphism described here will not be informative, since she is a homozygote for the 1.8Kb allele, and thus the normal and abnormal chromosomes cannot be differentiated. Another RFLP present in this gene may however prove to be useful. This is one of the disadvantages of RFLP work, because a succession of enzymes may have to be tried before a useful one is found.

RFLP linkage and Lod scores

A rather more circumspect use of RFLPs in carrier and prenatal diagnosis is by linkage. Here an RFLP is inherited with the gene defect because of their close proximity; but the RFLP *may* become separated from the disease gene by crossing over (recombination) at meiosis. Tracing the RFLP in the family would then give incorrect answers as to the inheritance of the gene defect in some members. The distance between the RFLP and the defective gene is obviously of the utmost importance; the closer they are the less likely they are to become separated. A measure of this distance is called the recombination fraction. This is estimated by the Lod score, which is calculated by means of the binomial theorem. This theorem initially looks rather forbidding, but is straightforward and worth the trouble taken in its understanding. It states that if, of two alternatives, the chance of one event happening equals q and the other equals 1 where $p+q=1$, then in a series of n events the chance of having $(n-r)$ of one sort and (r) of the other is equal to:

$$p^{(n-r)} \times q^{(r)} \times \frac{n!}{(n-r)!(r)!}$$

(where 3! means $3 \times 2 \times 1$, and by definition 0! is equal to 1).

If there is *no linkage* between the RFLP and the disease gene one would expect half the offspring to demonstrate no recombination, and the other half to be recombinants. What is the chance of observing in say ten individuals none to be recombinants? Using the binomial formula the chance would be:

$$0.5^{10} \times 0.5^{0} \times \frac{10!}{10!0!}$$

$= 0.001$, that is to say a very small chance.

However, if there were *some linkage* then one might expect say 70 per cent non-recombinants with 30 per cent recombinants. So in this case what would be the chance of observing no recombinants? This chance is:

$$0.7^{10} \times 0.3^{0} \times \frac{10!}{10!\,0!}$$

$= 0.03$; that is to say, it is thirty times more likely.

Finally if there was *close linkage* it might be expected that only 10 per cent would be recombinants. Again what is the chance in this case of observing no recombinants? The chance here is:

$$0.9^{10} \times 0.1^{0} \times \frac{10!}{10!\,0!}$$

= 0.3487; that is, it is over three hundred times more likely than on the first supposition.

The Lod score is the log of these *relative* odds. In the present case 0.3487 divided by 0.001 gives a relative odds figure of 348.7. The $\log_{10}$ of this is 2.5. We would probably say that in this case some linkage is evident, but in practice linkage is not generally accepted unless it is greater than 3. Lod scores from several families can be added together, so that an even more accurate linkage estimate can be made.

Obviously careful note must be taken of the recombination fraction when carrier detection or prenatal diagnosis is undertaken. In 1985 Reeders *et al.* (1985) reported linkage of an RFLP detected by a probe called 3′HVR to the dominantly inherited Adult Polycystic Kidney disease (APKD). They suggested that the Lod score 'peaked' at around 5 per cent recombination frequency (θ the recombination fraction = 0.05). In other words, the RFLP marker and the mutant APKD gene would be inherited together 95 per cent of the time, but in 5 per cent of cases they would become separated at meiosis. This can cause problems, as is illustrated in Fig. 3.6. The mother II_2 has the disease, and has had three children by two marriages. Two of these children are also affected. However, they have inherited different allelic restriction fragments, indicated here by letters of the alphabet. The mother has alleles GJ, and, since her normal mother has alleles JL, she must have received the mutant gene associated with allele G from her deceased affected father. Her child III_1 has inherited M from his normal father, and G, which is tracking the gene, from his mother. He is, as would be predicted, affected. However the daughter III_3 is also affected, but does not have allele G. Instead she has allele J from her mother, as does her normal brother III_4. It seems likely that III_3 is demonstrating a cross-over event between the RFLP and the defective gene which has occurred during oocyte formation. The result is that absence of allele G in III_3 does *not* indicate absence of the defective gene.

3.4 The future

The most straightforward method of identifying a defective gene is by cloning the cDNA from mRNA. This depends on being able to recognize either the mRNA itself or its protein product; this is not usually possible. A less straightforward way is by attempting to bypass the gene product and use RFLPs instead. This may be called 'reverse' genetics, because one is attempting to recognize the gene before its product. Unfortunately this is very labour-intensive, since an anonymous DNA sequence must be able to recognize an RFLP that is sufficiently polymorphic in a given population that family members will be frequently 'informative'. The RFLP must also

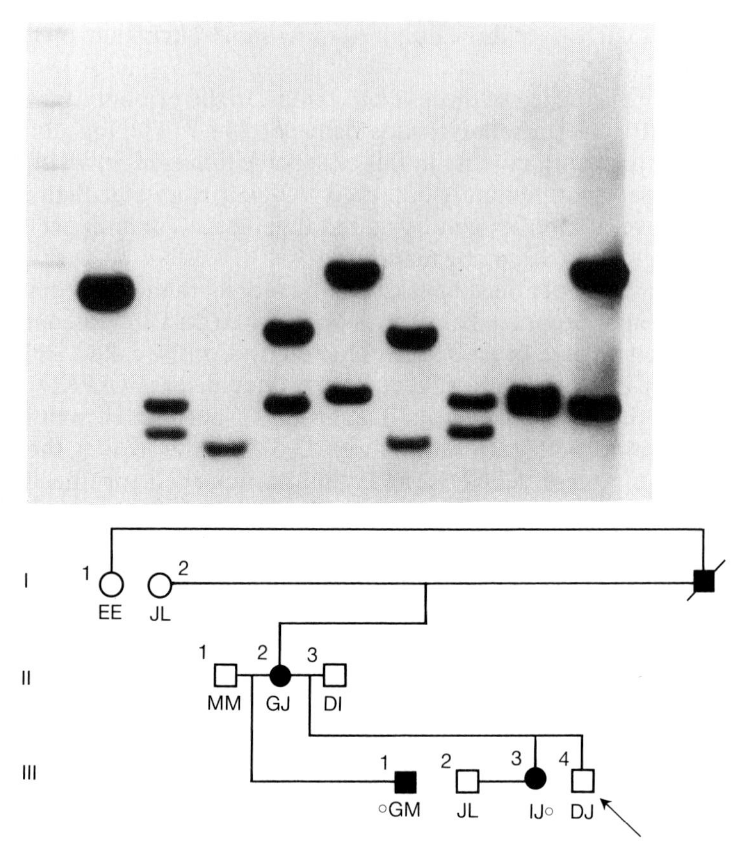

Fig. 3.6. Adult polycystic kidney disease pedigree and autoradiographic patterns.

be relatively closely linked to the disease gene, so that recombination between the two is infrequent.

Once a sequence fitting the bill has been cloned then predictive advice can be given. But the defect still has to be defined, since only then can therapy be envisaged. Furthermore, moving from even a closely linked cloned sequence to the gene is a time-consuming process. Thus to identify the defects involved in all of the four thousand or so genetic diseases will be a slow process. Some assistance may be forthcoming from the sequencing of the entire human genome, a process currently being undertaken, and which will take about three years to complete. Such a project will pin-point genes, but still require that such genes (in the defective state) are matched with diseases. Given that a gene defect is identified for a particular disease, what are the chances for a cure? It must be said that for most the chances are not good. Corrective surgery, diet control, and gene-product replacement therapy will probably continue to be used but more frequently.

Gene-replacement therapy has been discussed, but often with a lack of realization of the problems involved. To an extent it is already possible to insert genes into cells such that they are able to express themselves. However, genes must be place in particular cells, and with their regulatory functions intact, so that they are transcribed in the correct tissue and at the correct time. Both the inserted gene and the ones already present must be under control and must not adversely affect each other. Although recent work suggests some cause for optimism in this regard, there is still a long way to go yet!

References

Choo, K. H., Gould, G. K., Rees, D. J. G., and Brownlee, G. G. (1982). Molecular cloning of the gene for human auto-haemophilic factor IX. *Nature*, **299**, 178–80.

Giannelli, F., Anson, D. S., Choo, K. H., Rees, D. J. G., Winship, P. R., Ferrari, N., *et al.* (1984). Characterisation and use of an intragenic polymorphism for detection of carriers of haemophilia B (factor IX deficiency). *Lancet*, **i**, 239–41.

Kan, Y. W. and Dozy, A. M. (1978). Antenatal diagnosis of sickle cell anaemia by DNA analysis of amniotic fluid cells. *Lancet*, **ii**, 910–12.

Quaife, R. and Liu, D. T. Y. (1987). Extraction of DNA from chorionic villi. In *Chorion villus sampling* (ed. D. T. Y. Liu, E. M. Symonds, and M. S. Golbus), pp. 325–31. Chapman and Hall Medical, London.

Reeders, S. T., Breuning, M. H., Davies, K. E., Nicholls, R. D., Jarman, A. P., Higgs, D. R., *et al.* (1985). *Nature*, **317**, 542–4.

Saiki, R. K., Scharf, S., Faloona, F., Mullis, K. B., Horn, G. T., Erlich, H. A., *et al.* (1985). Enzymatic amplification of β-globin genomic sequences and restriction site analysis for diagnosis of sickle cell anaemia. *Science*, **230**, 1350–4.

4 Cytogenetic analysis from chorionic villi

Pat Cooke and Gerardine Turnbull-Ross

4.1 Introduction

Cytogenetic analysis can only be carried out on cells at the metaphase stage of division, when the chromosomes lie on a flat plane in the cell, and when they are condensed enough to be seen separately. All cytogenetic preparative techniques are therefore aimed at producing large numbers of cells in this stage. The cytogeneticist can then examine the number and structure of the sets of chromosomes. The objective of the analysis is to identify any numerical or structural deviation from normality in the chromosome set.

Unlike amniotic fluid cells, which have to be cultured to produce division, this process is occurring spontaneously in chorionic villus material. However, the rate of division is not homogeneous. Some fronds may be dividing vigorously, yielding adequate numbers of suitable cells for analysis, others may show no spontaneous divisions at all. Cytogenetic laboratories therefore apply several different preparative methods to each specimen in order to maximize the chance of obtaining analysable cells.

4.2 Providing your laboratory with a suitable sample

The first step in obtaining a successful result is the production of an adequate sample of material.

1. Villi should be collected into culture medium, supplemented with heparin (5 IU per ml).
2. The sample should be washed and dissected free of non-villous material in clean, heparinized medium.
3. The sample size should be checked. A minimum sample size of 10 mg wet weight has been suggested to be adequate for cytogenetic analysis (Fig. 4.1).

Fig. 4.1. Chorionic villus tissue—minimum samples size 10 mg wet weight.

4. The sample should be sent to the laboratory in clean medium supplemented with 15 per cent of calf serum, as quickly as possible. For successful direct preparations, a transfer time of up to 4 hours is acceptable. Longer delays will mean that some type of culture technique will have to be applied to revive division. Optimally, the transfer time for culture should be no longer than 24 hours.

4.3 Methods of preparation

The following sections describe the available preparative methods.

The direct method

Although chorionic villus material had been used for direct fetal sexing as early as 1975 (Anshan Department of Obstetrics and Gynaecology 1975) it was not until 1983 that Simoni and his colleagues (Simoni *et al.* 1983) reported on a reliable method for direct chromosome analysis from this material. A variety of techniques have now been reported, but all have the following characteristics in common.

1. Fronds with copious buds, if present, are selected for direct analysis, because they appear to produce better direct preparations than the flattened, stringy fronds which may also appear in the sample (Upadhyaya *et al.* 1987).
2. The material is incubated for between 1 and 2 hours in medium to which Colcemid has been added. This arrests divisions in progress at the metaphase stage.
3. A hypotonic solution is added, either to intact villi or to villus cells dispersed into a cell suspension by the use of a suitable enzyme, and the material is incubated for 15 to 20 minutes. This swells the cells so that the chromosomes become better separated.
4. The material is fixed, using a 3:1 mixture of methanol:acetic acid, and stored at −20°C for at least an hour.
5. Fixed cell suspensions are dropped on to clean slides and spread by surface tension. If complete villi are used, the cells are dispersed directly on to the slides using 70 per cent acetic acid, and spread mechanically.
6. The slides are then stained and banded.

Culture methods

A method of culturing villi for the production of chromosome preparations was first described by Niazi and her colleagues (1981). The early methods were complicated by the fact that fetal material did not respond well to conventional culture media, and also by difficulties associated with overgrowth of maternal tissue derived from adherent fragments of decidua or entometrium. The introduction of the use of a hormone-supplemented medium (Chang *et al.* 1982), which favoured the growth of fetal material, together with the realization that more stringent dissection of villi free from maternally derived tissue was necessary, minimized these problems. Culture methods today broadly follow the following protocol:

1. The villus fragments are carefully dissected free of material which may be maternally derived.
2. Fragments of villi, or cell suspensions prepared from them, are incubated in flasks for 24 hours in conventional medium (short culture) or are cultured with Chang medium over a longer period (long-term culture).
3. The processes of metaphase arrest, swelling, fixation, and slide-making follow the direct protocol.

In the early stages of the development of a service, when samples are few and far between, it is relatively easy to apply the full range of techniques to each specimen. As the number of specimens grows there is an incentive to try to minimize the amount of preparative work done. Most laboratories have carried out experiments to try to establish whether any single technique can reliably produce good results.

A good result is one where a reasonable number of metaphases are present (of the order of 5 per slide) in which the detailed banded structure of the chromosomes can be clearly seen. Figures 4.2 to 4.4 show metaphases of varying quality—from a poor metaphase in Fig. 4.2, where no banding has been obtained (solid stained) to a good metaphase in Fig. 4.4, where details of the banding pattern of each chromosome can be seen. Most reports suggest that the quality of banding is better in preparations from culture, rather than in direct preparations.

A recent paper (Holms *et al.* 1988) has crystallized an impression, generally held, about the relationship between method and quality, that is, that the greatest source of variation in the quality of preparations lies in differences between individual fronds within a specimen. Since the most appropriate method for any particular piece of material cannot be accurately predicted, it remains advisable to use a range of methods on each specimen to get the best results.

4.4 Timing of reports

This creates problems in advising patients on when they can expect a result. For some patients the direct method will provide an adequate number of

Figs 4.2–4.4. Metaphases of varying quality.

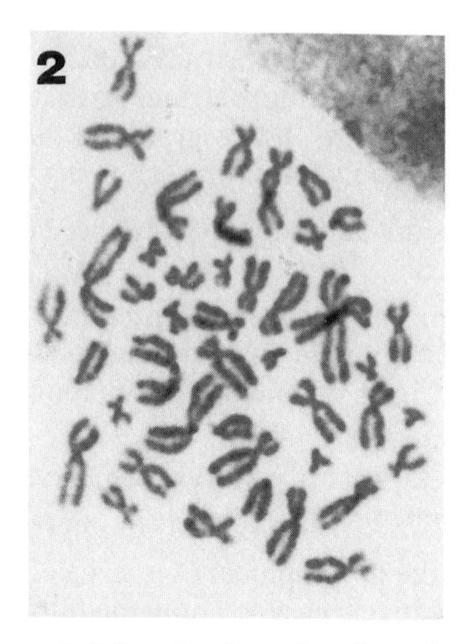

Fig. 4.2. Solid-stained—no banding obtained.

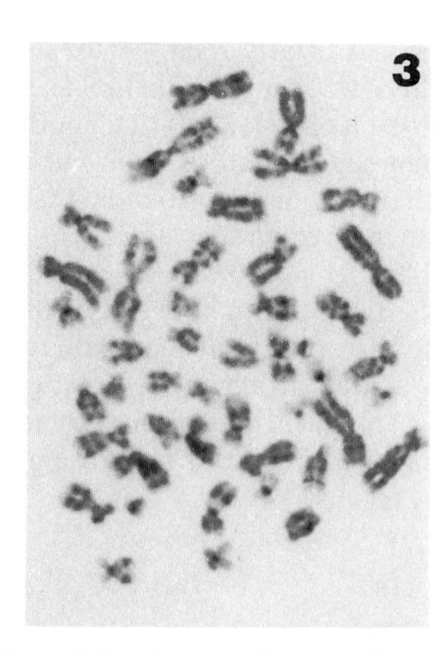

Fig. 4.3. Poor G-banding—only major bands visible.

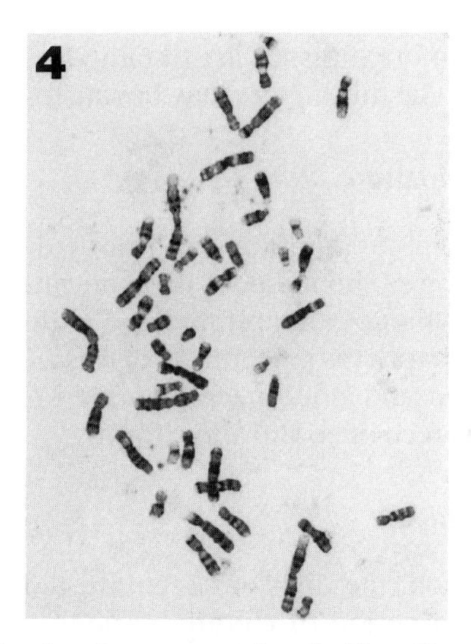

Fig. 4.4. Good G-banding—more detailed banding pattern visible.

metaphases of sufficient quality to give a definitive analysis—that is, an analysis on a large number of cells of sufficient banding quality to allow for a detailed check of the structure of each chromosome. They will receive a result within a few days of the sampling procedure. For others, the direct and short-term culture preparations will not yield enough metaphases, or will yield metaphases of poor banding quality. A definitive result must be obtained from cultured material, and so may not be available for over two weeks.

In most cases, the direct method produces at least a few metaphases, suitable for solid staining, which can therefore be checked quickly to establish the presence or absence of *major* numerical or structural abnormalities. However, such an analysis could miss the presence of mosaicism (a mixture of more than one cell type) together with some balanced and unbalanced structural chromosome changes. An obstetrician may choose to give a patient this provisional result on the clear understanding that, in a small proportion of cases, more detailed analysis from cultured material may reveal an undetected problem.

4.5 Diagnostic problems

However, there is another feature of cytogenetic results from direct versus cultured material which suggests that the wiser course of action may be to wait until the results of both analyses are available, because, in a small proportion of cases, different results are obtained from preparations made by different methods. The differences may be due to:

Maternal cell contamination

It is unlikely that a direct preparation will pick up divisions from maternal tissue, because of its lower spontaneous division-rate. However, in longer-term culture, maternal tissue can be stimulated to divide and contribute to the total number of metaphases present. As discussed earlier, this phenomenon is well understood, and modern methods have tended to eliminate this problem, or at least to recognize and allow for it.

Culture-induced artefacts

Using the direct method, the divisions seen are those which were taking place *in situ*. During a period of culture, accidents of division may take place because of the artificial environment for growth provided by even the best media and conditions.

Confined chorionic mosaicism

There may be variation between the chromosome complements of different villi. The differing results may represent genuine chromosomal heterogeneity within, and confined to, the chorion. Separate direct preparations, from individual fronds, can allow the recognition of this source of variation.

Variation within villi

Direct chromosome preparations reveal divisions taking place in the middle layer of the villus, the cytotrophoblast. In culture, most of the cells which grow and divide are fibroblasts from the mesodermal inner cell mass. Differing results, between direct and cultured material, may therefore reflect the different origin of the cells studied.

4.6 The interpretation of results

The interpretation of such conflicting results depends on the assessment, firstly, of whether the discrepancy is associated with contamination or culture artefacts, and, secondly, of which of the conflicting results best represent the chromosome complement of the fetus itself. This problem is an aspect of the general problem, common to prenatal diagnosis by both amniotic fluid cell-culture and chorionic villus analysis, that the sampled material may not represent the true status of the fetus (Kalousek and Dill 1983). The sampled material may itself be homogeneous, but different from that of the fetus; or the sampled material may be heterogeneous, with only one of its components representative of the actual fetal complement.

The fact that chorionic villus sampling takes place at an early stage of pregnancy offers the possibility of resolving ambiguous results by later analysis of amniotic fluid cells. However, in an attempt to minimize the number of cases where this is necessary, laboratories world-wide have been collecting data on discordant cases. From these results it was hoped that empirical rules, such as those available for the interpretation of ambiguous amniotic fluid cell-culture results, could be derived.

Tables 4.1 and 4.2 list published results, where analyses on direct, cultured, and fetal material were all available. Table 4.1 shows cases where there was discordance between the direct and the culture results. In these cases results from culture appear to be more reliable. The culture result was representative of the fetal karyotype in 22 of the 24 cases, while the direct analysis was representative in only 2. Table 4.2 shows three cases where there was agreement between the direct and culture results, showing the presence of a proportion of abnormal cells, but where, from fetal tissue, only normal cells were recovered.

From these published data it is apparent that there is still ambiguity about the interpretation of results when direct and cultured results conflict, and, even when these results agree, where mosaicism is present. In these cases amniocentesis should be recommended to resolve the uncertainty.

4.7 Reliability of cytogenetic analysis from villi

It is very difficult indeed to estimate the frequency with which diagnostic problems may occur. Even the precise frequency of failure to obtain a chromosome result is complicated by the fact that some reports include partial or complete sampling failure in their total failure-rate, whereas others exclude cases where an inadequate sample was obtained before calculating their failure-rate for chromosome preparation alone. As an approximation, in experienced hands the aggregate failure-rate (including that associated with inadequate samples) should not exceed 5 per cent.

Similarly, it is difficult to obtain precise figures on the frequency of diagnostic problems following 'successful' chromosome analysis. Most problem cases are reported individually, with no indication of the total number of cases from which the problems were identified. Even when surveys of large numbers of cases are described, many of the pregnancies have not been completed at the time of the report, and so there is a chance that some unsuspected discrepancies may arise when the whole set of

Table 4.1. Cases of discordance. The results of analysis from direct and cultured material, compared with the fetal karyotype

Source	Direct	Culture	Fetus
Wapner *et al.* 1985	46,XX/47,XX, + 13	46,XX	46,XX
	46,XY/47,XY, + 13	46,XY	46,XY
	46,XX/47,XXX	46,XX	46,XX
Simoni *et al.* 1985	46,XX/47,XX, + 3	46,XX	46,XX
	46,XY/45,X	46,XY	46,XY
Linton and Lilford 1986	46,XY	47,XXY	46,XY/47,XXY
Eichenbaum *et al.* 1986	46,XY	46,XY/47,XXY	46,XY/47,XXY
Martin *et al.* 1986	46,XY	47,XY, + 18	47,XY, + 18
Kalousek *et al.* 1987	47,XY, + 12	46,XY	46,XY
	47,XX, + mar	46,XX	46,XX
Schulze *et al.* 1987	46,XX/47,XX, + 3	46,XX	46,XX
Verjaal *et al.* 1987	46,XYq +	46,XY	46,XY
	46,XX/47,XX, + 12	46,XX	46,XX
Therkelsen *et al.* 1988	46,XX,t(6;22)/46,XX	46,XX	46,XX
	47,XY, + 20/46,XY	46,XY	46,XY
	45,X/46,XX	46,XX	46,XX
	46,XY	47,XY, + 5/46,XY	46,XY
	47,XXY/46,XY	46,XY	46,XY
Crane and Cheung 1988	46,XX/45,X	46,XX	46,XX
	46,XY/47,XY, + mar	46,XY	46,XY
	46,XY/47,XXY	46,XY	46,XY
	46,XX/47,XX, + 13	46,XX	46,XX
	46,XY/47,XY, + 21	47,XY, + 21	47,XY, + 21
	46,XX	46,XX/47,XXX	46,XX

Table 4.2. Cases with agreement between direct and cultured material but discordance with fetal karyotype

Source	Direct	Culture	Fetus
Simoni *et al.* 1985	46,XY/47,XY, + 18	46,XY/47,XY, + 18	46,XY
Schulze *et al.* 1987	46,XX/47,XX, + 15	47,XX, + 15/46,XX	46,XX
Therkelsen *et al.* 1988	47,XY, + mar/46,XY	47,XY, + mar/46,XY	46,XY

deliveries have been completed. In recognition of the need for more precise data, National Collaborative Studies have been set up in most countries.

4.8 Follow-up of tested pregnancies

The most comprehensive report so far is that of Simoni *et al.* (1985) who gave results from 1000 samples from pregnancies of which 677 had, at that time, gone to term. Failure to obtain chromosome analysis at the first attempt, from adequate samples, occurred in only 1.4 per cent of cases. Forty-seven numerical, 2 unbalanced strutural, and 21 balanced structural abnormalities were detected. Of these, the karyotype was confirmed as correct in 31 cases. Confirmation was not available or reported in 31 cases (of which 22 were cases with balanced rearrangements).

Discrepancies were reported in 8 cases, of which 7 cases showed mosaicism in villus material which was not confirmed at amniocentesis (2 cases) or at termination (5 cases). With current methods, all of these results would be checked by amniocentesis, avoiding a false-positive result. In only one case, where an unbalanced structural abnormality was found in all the villus cells examined, would a false-positive result be given now.

On the assumption that all unconfirmed results and all pregnancies in progress would reveal no further anomalies, and given current methods, where both direct and culture results are obtained and discrepancies and mosaics are investigated further through amniocentesis, this series would represent a misdiagnosis rate of only 0.1 per cent.

Any cytogenetics laboratory which carries out tests on chorionic villus specimens should obtain follow-up on the outcome of all tested pregnancies.

1. Where the chromosomal result is normal, with no analytical problems, it is probably sufficient to record whether or not the baby was phenotypically normal, of appropriate sex.
2. Where an abnormality has been found which has led to termination of pregnancy, confirmation of the cytogenetic diagnosis should be

obtained by culture from fetal material. Unfortunately, some obstetricians use a method of termination which interferes with culture growth of fetal material, and so this is not always possible.

3. Where a balanced abnormality has been detected, particularly where a female fetus and mother apparently carry the same rearrangement, and the pregnancy has been allowed to continue to term, a phenotypically normal outcome of appropriate sex should still be checked, preferably by chromosome analysis of the baby's blood at birth.
4. When an ambiguous result has been obtained (for example where direct and culture results conflict, or where mosaicism has been present), and even when amniocentesis appears to have resolved the difficulty and the pregnancy continues to term, it is particularly important to check the outcome of the pregnancy so that we can build up a body of hard data which will help in the task of accurate future interpretation. Preferably, this should be done by chromosome analysis of the baby's blood at birth.

Unfortunately in the latter cases (3 and 4), having reassured the parents that a phenotypically normal outcome is to be expected, it can be difficult to introduce the suggestion that the chromosome complement of the baby should be further checked. Also, we know that the examination of the chromosome complement of the baby's blood alone cannot exclude the possibility of the presence of a different cell line in another tissue. Follow-up can also be difficult in the rare cases where miscarriage occurs or where the family moves, between the time of the test and delivery, to a different area. Nevertheless, it is only by the careful compilation of the outcomes of tested pregnancies that the problems and limitations of the test system can be established and resolved.

4.9 Experience in Nottingham

From our limited experience at the time of writing, where we have studied 344 pregnancies (of which 242 have been completed, with follow-up available on 228, and with 102 pregnancies continuing) no known diagnostic errors have occurred. Our success-rate over the last two years has been 97 per cent. About half these failures have been associated with inadequate samples. The proportion of cases where results from both direct and culture analyses are available has risen from 2 per cent to 65 per cent.

Of the 228 completed pregnancies which have been followed up, 213 had normal karyotypes and phenotypically normal outcomes of appropriate sex. Three were chromosomally normal, with an outcome of appropriate sex, but were terminated because of abnormal findings on single-gene DNA analysis. Three confirmed cases were found to carry balanced rearrangements, inherited from one or other parent. Five carried unbalanced chromosome complements, leading to termination of pregnancy and confirmation of the abnormal result by the analysis of fetal tissue. There were four ambiguous results.

Case 1. One direct preparation, from a single frond, gave the karyotype 47,XX,+iso(11q). All the other fronds gave normal female karyotypes. This was reported as a normal female result, and has been confirmed as such at term.

Case 2. A number of fronds showed mosaicism, 47,XY,+7/46,XY. Amniocentesis was recommended, although a normal male result was anticipated. Nevertheless, the patient opted for termination, and only 46,XY cells were seen in fetal tissue.

Case 3. Direct preparations showed 47,XY,+12. Cultured villi gave a 46,XY karyotype. Amniocentesis was recommended, and a normal 46,XY result obtained. The pregnancy resulted in the birth of a phenotypically normal male.

Case 4. Direct preparations gave a 45,X result. Cultured material was uniformly 46,XX. Amniocentesis was recommended, and a mosaic 45,X/46,XX result obtained. The method of termination precluded analysis of fetal material.

From this limited data-set, problems of interpretation presented in less than 2 per cent of cases. However, these problems could be resolved by the combination of analyses from direct preparations from individual fronds and from culture, together with follow-up by amniocentesis in discordant cases. Of the 228 cases where the outcome is known, reassurance of an expected normal pregnancy outcome was correctly given to women at high risk in almost 95 per cent of cases. In eight cases, the opportunity was given for early termination of an abnormal fetus.

4.10 Conclusion

The overwhelming advantages which are offered to the patient by chorionic villus sampling, rather than amniocentesis, appear to outweigh the potential problems of interpretation. The principal task of prenatal diagnosis, that of assuaging the anxiety of families at risk of genetic disorder, is carried out early in pregnancy. Where abnormality is found, early termination is less disruptive to the family, and safer and less traumatic for the mother.

References

Anshan Department of Obstetrics and Gynaecology (1975). Fetal sex prediction by sex chromatin of chorionic villi cells during early pregnancy. *Chinese Medical Journal* (English version), **1**, 117–26.

Chang, H. C., Jones, O. W., and Masui, H. (1982). Human amniotic fluid cells grown in a hormone supplemented medium: suitability for prenatal diagnosis. *Proceedings of the National Academy of Sciences* (USA), **79**, 4795–9.

Crane, J. P. and Cheung, S. W. (1988). An embryonic model to explain cytogenetic inconsistencies observed in chorionic villus versus fetal tissue. *Prenatal Diagnosis*, **8**, 119–29.

Eichenbaum, S. Z., Krumins, E. J., Fortune, D. W., and Duke, J. (1986). False negative finding on chorionic villus sampling. *Lancet*, **ii**, 391.

Holmes, D. S., Fifer, A. M., Mackenzie, W. E., Griffiths, M. J., and Newton, J. R. (1988). Direct and short-term culture preparation of chorionic villi. Is any one method best? *Prenatal Diagnosis*, **8**, 501–9.

Kaousek, D. K. and Dill, F. J. (1983). Chromosomal mosaicism confined to the placenta in human conceptions. *Science*, **221**, 665–7.

Kalousek, D. K., Dill, F. J., Pantzar, T., McGillivray, B. C., Li Yong, S., and Douglas-Wilson, R. (1987). Confined chorionic mosaicism in prenatal diagnosis. *Human Genetics*, **77**, 163–7.

Linton, G. and Lilford, R. J. (1986). False-negative finding on chorionic villus sampling. *Lancet*, **ii**, 630.

Martin, A. O., Elias, S., Rosinski, B., Bombard, A. T., and Simpson, J. L. (1986). False-negative finding on chorionic villus sampling. *Lancet*, **ii**, 391.

Niazi, M., Coleman, D. V., and Loeffler, F. E. (1981). Trophoblast sampling in early pregnancy. Culture of rapidly dividing cells from immature placental villi. *British Journal of Obstetrics and Gynaecology*, **88**, 1081–5.

Schulze, B., Schlesinger, C., and Miller, K. (1987). Chromosomal mosaicism confined to chorionic tissue. *Prenatal Diagnosis*, **7**, 451–3.

Simoni, G., Brambati, B., Danesino, C., Rossella, F., Terzoli, G. L., Ferrari, M., *et al.* (1983). Efficient direct chromosome analyses and enzyme determinations from chorionic villi samples in the first trimester of pregnancy. *Human Genetics*, **63**, 349–57.

Simoni, G., Gimelli, G., Cuoco, C., Terzolli, G. L., Rossella, F., Rometti, L., *et al.* (1985). Discordance between prenatal cytogenetic diagnosis after chorionic villus sampling and chromosomal constitution of the fetus. In *First trimester fetal diagnosis* (ed. M. Fraccaro, G. Simoni, and B. Brambati), pp. 137–43. Springer-Verlag, Heidelberg.

Therkelsen, A. J., Jensen, P. K. A., Hertz, J. M., Smidt-Jensen, S., and Hahnemann, N. (1988). Prenatal diagnosis after transabdominal chorionic villus sampling in the first trimester. *Prenatal Diagnosis*, **8**, 19–31.

Upadhyaya, M., Jansani, B., Little, E., Harper, P. S., Rees, D., and Roberts, A. (1987). Lack of sampling site variation in chorionic villus biopsy as assessed by DNA, enzymatic, morphological and cytogenetic analysis. *Prenatal Diagnosis*, **7**, 119–27.

Verjaal, M., Leschot, N. J., Wolf, H., and Treffers, P. E. (1987). Karyotypic differences between cells from placenta and other fetal tissues. *Prenatal Diagnosis*, **7**, 343–8.

Wapner, R., Jackson, L., Davis, G., Barr, M. and Hux, C. (1985). Cytogenetic discrepancies found at chorionic villus sampling (CVS). *American Journal of Human Genetics* (July supplement), **37**(4), A122.

5 The interpretation of cytogenetic results

Denise E. Rooney

5.1 Introduction

The prenatal diagnosis of chromosome defects is dependent upon the assumption that the tissue being analysed has the same karyotype as the fetus. In the majority of cases this assumption is valid, but anyone familiar with amniocentesis will already be aware that exceptions do occur. The interpretation of mosaicism and pseudomosaicism in amniocentesis has been widely debated in the literature, and a number of useful guidelines exist to aid in the management of difficult cases (Hsu and Perlis 1984; Worton and Stern 1984; Bui *et al.* 1984).

It was, perhaps, not surprising therefore that similar problems were likely to occur with the use of chorionic villus sampling (CVS) when this was established as an alternative method of prenatal diagnosis. However, the problems of false diagnosis and mosaicism in CVS and their interpretation are somewhat different from the situation in amniocentesis. In this chapter, we shall consider how the karyotype of the chorionic villi may differ from that of the fetus, how we can be alerted to the possibility of a false diagnosis, and how to interpret an equivocal result correctly.

5.2 How can the karyotype of the chorionic villi differ from that of the fetus?

Confined placental mosaicism

After fertilization and early cleavage, the 16 cells forming the morula divide and differentiate to form the blastocyst. The majority of these cells will form the cytotrophoblast, which becomes the outer layer of the chorionic villi. Only a few cells will form the inner cell mass, and of these, only three or four will form the actual embryo.

When an abnormal cell-division occurs after fertilization, the distribution of cells throughout the embryonic tissues bearing an abnormal karyotype will depend on the stage of development at which that division occurred. Thus an abnormal division occurring before the morula stage is most likely to produce an overall mosaicism present in both placental and fetal tissues. A later aberrant division will lead to abnormalities being confined to the tissues which have been formed by the particular cells deriving from that division. For example, an abnormal cell-division arising in a cytotrophoblast cell will produce mosaicism or abnormality confined to the placenta, whereas the embryo itself would have a normal karyotype. This situation could give rise to a *false positive* result. Conversely, a mis-division in a cell of the inner cell mass destined to form the embryo will produce a mosaic or abnormal fetus with a normal placenta, and thus could give rise to a *false negative* result.

Since more cells are destined to become cytotrophoblast (that is, placenta) than inner cell mass (fetus), there is a greater likelihood of mosaicism occurring in the placenta than in the fetus. This is reflected in the observed preponderance of false positive results over false negative results.

This is the basic model of confined placental mosaicism proposed by Kalousek and Dill (1983) and Kalousek (1985). Crane and Cheung (1988) extended this model to explain the growing number of reports of discrepancies between karyotypes of different cell types within the chorionic villus itself.

Discrepancies between the karyotypes obtained from cell culture and direct preparations

Chapter 4 describes how chromosome preparations can be made either directly from uncultured chorionic villi, or from cells obtained from tissue culture. The two methods utilize different cell types within the chorionic villi. The spontaneously dividing cells exploited by the direct preparations are to be found in the cytotrophoblast layer, whereas the tissue cultures represent mainly cells derived from the mesenchyme core. In the early blastocyst, the mesenchyme core is derived from cells in the inner cell mass.

An aberrant division in the cytotrophoblast may thus produce abnormal cells represented only in that layer of the chorionic villus. Conversely, an aberrant division in the cells of the inner cell mass may lead to mosaicism or abnormality confined to the mesenchyme core.

A mis-division in the inner cell mass may, of course, lead to mosaicism or abnormality in both mesenchyme core and embryo, although the cytotrophoblast would be normal.

Summary of possible discrepant situations

False positive:

(1) Mosaic or abnormal placenta, normal fetus;

(2) mosaic or abnormal direct preparations, normal cultures and fetus; and

(3) mosaic or abnormal culture, normal direct preparations and fetus.

False negative:

(4) Normal placenta, mosaic or abnormal fetus;

(5) normal direct preparations, mosaic or abnormal cultures and fetus; and

(6) normal cultures, mosaic or abnormal direct preparations and fetus.

Other sources of error

Disappearing twin

Although the above models for discrepant karyotypes probably explain the majority of cases, it is likely that a few problems arise as a result of a failed twin pregnancy. The 'vanishing twin' theory postulates the existence of a chromosomally abnormal twin sac which fails to develop, and is eventually reabsorbed but for some persisting chorionic villi. An example of this is reported by Tharapel *et al.* (1989), who detected cells trisomic for chromosome 16 confined to villi obtained from a placental nodule detected on ultrasound.

Localized placental mosaicism

In the same way that an aberrant cell division can lead to confined placental mosaicism, abnormal cells can be localized to a small area within the placenta. This would result from a mis-division at a late stage of placental development. Consequently, in a small minority of cases, it would be possible to sample villi with an abnormal karyotype which is not representative of the placental karyotype. This sometimes becomes apparent when villi from separate samplings from the same pregnancy are processed independently of one another, and show different karyotypes.

Maternal cell contamination

The chorionic villi are closely associated with the uterine decidua, and aspirates often contain a mixture of the two tissues. In some cases the decidual fragments may be difficult to separate from the chorionic villi. Furthermore, the sample may contain endometrial gland tissue, which can superficially resemble the chorionic villi.

If an unsorted mixture of these tissues is cultured, the cell growth will reflect both fetal and maternal karyotypes. Clearly the villi must be separated from the maternal tissues by an experienced cytogeneticist before they are cultured. Thus the occurrence of maternal cell contamination will depend upon the skill of the tissue culturist. In some cases, however, there may be maternal cells present in the villus sample that are not visible even to the most experienced operator, and these in a very few cases may lead to a mixed karyotype.

The direct method, on the other hand, depends entirely upon the spontaneously dividing cells of the cytotrophoblast, and most workers agree that maternal cell contamination is unlikely to arise in these preparations.

Undetected chromosome rearrangements

Straightforward numerical chromosome abnormalities are, on the whole, simple to detect, and do not require a high level of banding resolution. The detection of more subtle chromosome rearrangements is dependent on chromosome morphology and banding quality. It is generally acknowledged that chromosome preparations obtained from blood-samples are superior, in most cases, to those obtained from amniotic fluid cell-cultures.

This means that fewer abnormalities can be detected by amniocentesis than by blood-culture. However, since the population being screened for prenatal diagnosis is at a comparatively low risk for *de novo* subtle chromosome abnormalities, the sensitivity of amniocentesis is considered to be acceptable. The same is generally true for CVS, but many laboratories find that cultured villi yield chromosomes of a slightly inferior quality compared to those from amniotic fluid cultures. Preparations are, however, usually of sufficient quality to exclude most chromosome rearrangements at the ISCN 500 band level.

The direct method yields preparations which are variable both in mitotic index and chromosome morphology. It is also difficult to obtain banding of a quality sufficient to exclude any more than numerical and large chromosome rearrangements visible at the ISCN 100 band level.

It is therefore possible, in rare cases, to miss a subtle chromosome abnormality because of the insufficient quality of the preparation. This is more likely to occur with direct preparations.

Fragile X expression

Techniques are available to promote expression of Fragile X in CVS cultures. However, these are not entirely reliable, and failure to express Fragile X in the CVS cultures of a fetus at risk does not exclude this condition.

5.3 How can we be alerted to the possibility of a false diagnosis?

Analysis of both direct and cultured preparations

Having examined the ways in which discrepant results can occur, it is perhaps clear that an analysis of cells obtained from both cultured and direct preparations could, in some cases, highlight potential problems.

From the summary of possible discrepant situations shown above it can be seen that analysis of direct preparations alone would give a false result in situations 2 and 5, whereas analysis of cultures alone would give a false result in situations 3 and 6. There are examples of each of these situations in the literature, with disastrous consequences for the patients and their pregnancies.

Analysis of direct and cultured preparations together can also reveal maternal cell contamination. It is likely that if a female karyotype is detected in a culture, but the direct preparations show a male karyotype, maternal cell contamination is present. Maternal cell contamination should be suspected in a culture where both 46,XX and 46,XY cells are detected.

5.4 Interpretation of equivocal results

The interpretation of abnormal results depends on three factors:
(1) The patterns of mosaicism if present;
(2) results obtained from direct and cultured preparations; and
(3) the chromosome abnormality involved.

Levels of mosaicism

Much has been published about the interpretation of mosaicism in amniotic fluid cultures, and different levels of mosaicism have been defined thus:

Level 1 mosaicism: single-cell abnormality.

Level 2 mosaicism: multiple cells with the same abnormality in a single flask or cell colony.

Level 3 mosaicism: multiple cells with the same abnormality distributed over multiple flasks or cell colonies.

Levels 1 and 2 mosaicism are not normally representative of fetal karyotype, whereas level 3 mosaicism is usually genuine. Likewise, single-cell anomalies in CVS preparations can, in the majority of cases, be ignored. Levels 2 and 3 mosaicism relate only to cultured CVS preparations, and assume that corresponding direct preparations are normal. Interpretation of these situations will depend mostly on the chromosome abnormality involved.

Results obtained from direct and cultured preparations

The importance of comparing results obtained from direct and cultured preparations has already been discussed. In general, an abnormal result obtained from both direct and cultured preparations is likely to represent genuine fetal abnormality as long as the chromosome involved is consistent with a viable pregnancy.

Any discrepancy should alert the clinician to the possibility of a false result, especially if the abnormality appears in mosaic form in one preparation only. In general, false positive results are more frequent in direct preparations than in cultures.

It is difficult to determine the exact risk of false results, but the estimates most often quoted are:

False positive: 1 per cent;
False negative: 0.1 per cent.

These figures are based on a large Italian collaborative study (Simoni *et al.*

1987), but other collaborative studies (Canadian Collaborative CVS–Amniocentesis Clinical Trial Group 1989) have suggested that the overall false positive rate may be as high as 2 per cent.

Which chromosome is involved?

Having applied the previous factors to the abnormality in question, it is essential to consider the implications of the chromosome abnormality that has been detected. This is probably the single most important factor to consider when interpreting an abnormal result. The chromosome abnormalities which can occur in CVS can be broadly divided into three categories according to their likely significance:

1. Abnormalities which are compatible with a full-term pregnancy, but which lead to phenotypic abnormality even in mosaic form. This includes trisomies of chromosomes 13, 18, and 21, triploidy, and sex-chromosome abnormalities. The detection of any of these in non-mosaic form in both direct and cultured preparations is a good indication of a genuine abnormality. If they are detected in mosaic form, or as a discrepancy between direct and cultured preparations, amniocentesis should be seriously considered.

2. Abnormalities which are incompatible with full-term pregnancy, but which in mosaic form may lead to delivery of an abnormal baby. This includes all other autosomal trisomies except those in category 3. The detection of these autosomal trisomies in non-mosaic form, even if they are present in both direct and cultured preparations, should be regarded with suspicion if the ultrasound scan shows a normally developing fetus. In these cases, it may not be necessary to proceed to any further investigations other than regular detailed scans throughout the pregnancy. If these trisomies occur as mosaics, or as discrepancies between the direct and cultured preparations, then amniocentesis should be seriously considered. Trisomy 8 mosaicism, in particular, should be thoroughly investigated.

3. Abnormalities which are incompatible with a viable pregnancy, even in mosaic form. This includes trisomies of chromosomes 2 and 16. Pseudomosaicism of chromosome 2 is very common in amniotic fluid cultures, whereas trisomy 16 is the most common autosomal trisomy found in early spontaneous abortion studies. There are no reported cases of any of these trisomies, or their mosaics, leading to a full-term pregnancy. Therefore this result, associated with an ultrasound scan showing a normally developing fetus, does not require further investigation, and the patient can be reassured that there is no cause for concern.

Tetraploidy

Tetraploid mosaicism is often found in both amniotic fluid and CVS cultures, and laboratories very rarely report this. It is also fairly common in direct CVS preparations. If full tetraploidy is detected in both direct and cultured preparations it is probably not necessary to offer amniocentesis unless abnormality is suspected from the ultrasound scan. Such a pregnancy should, however, be carefully monitored by scan. In practice, however, some laboratories do prefer to offer amniocentesis as a precaution.

Maternal cell contamination

If maternal cell contamination is suspected, it is usual to attempt to exclude genuine sex-chromosome anomaly by comparing the polymorphisms of the cultured cells with those in cells from maternal and paternal lymphocyte cultures. This does not always clarify the situation, especially if maternal and paternal polymorphisms are similar. In most cases, however, maternal cell contamination is the most likely explanation for a mixture of male and female karyotypes, and there is very little necessity for further investigation.

Structural rearrangements

Structural rearrangements detected by CVS can, for the most part, be interpreted in the same way as those detected by amniocentesis. This will usually involve examination of parental bloods for the rearrangement in question, and risks of *de novo*, apparently balanced rearrangements leading to phenotypic abnormality are normally in the region of 5–10 per cent, depending on the number of chromosomes involved. However, a *de novo* rearrangement present as a discrepancy in either direct or cultured preparations is less likely to be genuine, but may still warrant confirmatory tests, depending on the nature of the anomaly.

Marker chromosomes

Small supernumerary marker chromosomes, often bisatellited, may be detected by prenatal diagnosis. Sometimes these are familial, and have no apparent effect on phenotype, but occasionally they may be associated with abnormality. These should be treated in a similar manner to structural rearrangements, and parental bloods should be examined. *De novo* marker chromosomes carry a similar risk to the phenotype as a *de novo* rearrangement. They do, however, also occur in CVS as confined placental mosaicism, and a *de novo* marker detected as a discrepancy between direct and cultured preparations should be investigated by amniocentesis.

Confined placental mosaicism and pregnancy outcome

Since the placenta is an organ vital to the pregnancy, it is possible that karyotypically abnormal cells might disturb the placental function. This possibility has been debated in the literature, and cases are reported of alleged association between confined placental mosaicism and intra-uterine growth retardation (IUGR) and spontaneous abortion. The evidence for an association in all cases of karyotypic abnormality is not yet strong, however, and more studies on the placentae of IUGR and abortions are needed.

5.5 The success-rate of karyotyping CVS

Most laboratories are capable of obtaining a result from at least 95 per cent of samples, and many claim success-rates of up to 98 per cent, which is similar to that obtained from amniocentesis. The success-rate will, of course, depend largely on the quality of the sample provided. It is therefore important that someone trained to assess the specimen is present to evaluate the sample immediately, so that further material can be obtained if necessary. This involves being able to recognize active villi, and to be able to distinguish them from maternal tissues. Although it is sometimes possible to obtain a result from a small quantity of villi, in reality the test requires at least 10 mg of good-quality material.

Direct preparations are possible within a few hours of sampling, although many laboratories prefer to use an overnight incubation procedure. Culture times vary, but preparations can normally be made from 1 week onwards. This will also depend on the quality and quantity of the sample provided. Reporting times will depend on work-load. It must be remembered that CVS preparations are extremely time-consuming, and an excessive work-load will result in long reporting times.

Follow-up studies following termination of pregnancy

It is important that, in the event of termination of pregnancy, fresh unfixed products of conception are sent to the laboratory for confirmation of karyotype, since it is from such studies that the accuracy of CVS results can be monitored.

5.6 Counselling the patient

Having considered the possible ways in which equivocal results can arise, it must be pointed out that the results are straightforward in at least 98 per cent of cases.

The greatest advantage of CVS over other prenatal tests is that it can be carried out in the first trimester, and that results can be obtained rapidly. This means that termination of pregnancy, if necessary, can be carried out within the first trimester by suction, and the traumatic procedures associated with later termination can be avoided. It was hoped, in the early days of CVS, that all results would be available within a few days based on the direct method of obtaining a karyotype. For the reasons already discussed, many cytogeneticists now discourage the use of direct procedures alone, and prefer to report from analyses based on both direct and cultured preparations. Conversely, many clinicians continue to offer the test on the basis of rapid results, which can be given to the patients within a few days of the test. The correct approach will vary to some extent, depending on the stage of gestation beyond which suction termination is unavailable; indeed, some countries do not allow termination at all beyond the first trimester of pregnancy. Thus the reporting policy for CVS has to take account of these situations.

In general, the ideal approach would be to present the test as a two-stage procedure, whereby the rapid result based on the direct method is regarded as a preliminary result, to be confirmed a week or two later by the final result based on culture.

It is important to be realistic when telling the patient how long she should have to wait for a result. Most laboratories will try to obtain a result within a day or two of sampling for cases at very high risk—for instance sex-linked disorders. On a practical level, however, most routine cases, such as maternal-age screens, may take a couple of days longer. An over-optimistic estimation of the time it will take to obtain a result will lead to a barrage of telephone calls and a disappointed patient!

The importance of honest and informed counselling cannot be stressed enough, for this can make the difference between a patient's terminating her pregnancy unnecessarily, or proceeding to have a normal, healthy baby. Only if the clinician is aware of the possible problems that can arise with CVS, and is armed with the information with which to interpret them, can the pregnancy be correctly managed.

References

Bui, T.-H., Iselius, L., and Lindsten, J. (1984). European Collaborative Study on Prenatal Diagnosis: mosaicism, pseudomosaicism and single abnormal cells in

amniotic fluid cultures. *Prenatal Diagnosis*, **4**, 145–62.
Canadian Collaborative CVS–Amniocentesis Clinical Trial Group (1989). Multicentre randomised clinical trial of chorion villus sampling and amniocentesis. *Lancet*, **i**, 1–6.
Crane, J. P. and Cheung, S. W. (1988). An embryonic model to explain cytogenic inconsistencies observed in chorionic villus versus fetal tissue. *Prenatal Diagnosis*, **8**, 119–29.
Hsu, L. Y. F. and Perlis, T. E. (1984). United States survey on chromosome mosaicism and pseudomosaicism in prenatal diagnosis. *Prenatal Diagnosis*, **4**, 97–130.
Kalousek, D. K. (1985). Mosaicism confined to chorionic tissue in human gestations. In *First trimester fetal diagnosis*, ed. M. Fraccaro, G. Simoni, and B. Brambati, 130–6. Springer-Verlag, Heidelberg.
Kalousek, D. K. and Dill, F. J. (1983). Chromosomal mosaicism confined to the placenta in human conceptions. *Science*, 221, 665–7.
Simoni, G., Fraccaro, M., Gimelli, G., Maggi, F., and Dagna Bricarelli, F. (1987). False-positive and false-negative findings on chorionic villus sampling. *Prenatal Diagnosis*, **7**, 671–2.
Tharapel, A. T., Elias, S., Shulman, L. P., Seely, L., Emerson, D. S., and Leigh Simpson, J. (1989). Resorbed co-twin as an explanation for discrepant chorionic villus results: non-mosaic 47, XX, +16 in villi (direct and culture) with normal (46, XX) amniotic fluid and neonatal blood. *Prenatal Diagnosis*, **9**, 467–72.
Worton, R. G. and Stern, R. (1984). Canadian Collaborative Study of Mosaicism in amniotic fluid cultures. *Prenatal Diagnosis*, **4**, 131–44.

6 Ultrasound in early pregnancy

P. Twining

6.1 Introduction

A short description of the ultrasonic appearances in the first trimester of pregnancy is presented. The chapter starts with a brief account of the physics of ultrasound, followed by a description of the machines used and the techniques of scanning. Ultrasound appearances and relevant embryology are illustrated, followed by some examples of abnormal appearances found in early pregnancy.

6.2 Physical principles of ultrasound

Basic physics

An ultrasound image is produced from high-frequency sound waves far above the audible range. The frequencies used in the diagnostic range are 3.5–5.0 MHz for abdominal scanning and 5.0–7.0 MHz for vaginal scanning. The frequency of the transducer or probe is important, as the higher the frequency the better the resolution or clarity of the image. There is a drawback, however, because as the frequency increases, the depth of penetration of ultrasound decreases: in other words, deep structures will not be seen with a high-frequency transducer.

In practice, a compromise is achieved, with a transducer that will produce a reasonably clear image to an adequate depth; hence most obstetric scanning is carried out using a 3.5 MHz transducer. Early pregnancies can be scanned using a 5 MHz transducer; but this is not suitable for late pregnancies. Vaginal scanning can use higher frequencies, as the uterus is much closer to the transducer.

The transducer is made up of a large number of piezo-electric crystals. These crystals emit high-frequency sound-waves in response to a rapidly changing voltage. The crystals are therefore capable of transforming electrical energy into sound-waves. The converse is also true, so that if high-frequency sound vibrates the crystals, a small voltage will be produced.

The principle of ultrasound scanning relies on a high-frequency pulse of sound being emitted from the crystal. This sound enters the body and travels through the tissues at 1540 metres per second. Some of the sound will be absorbed by the tissues, but some will be reflected back towards the crystal. These reflections usually occur at tissue interfaces, for example, between the bladder-wall and the uterus, or between the liquor and the fetus.

After the crystal has emitted the sound, it then 'listens' for any returning pulses or sound; this reflected sound vibrates the crystal, and a small electrical voltage is produced. This electrical voltage will be amplified to create a dot on the television screen, and this dot represents a tissue interface within the body.

Using a large number of crystals an image can be built up of the various tissue interfaces. The image that is displayed represents a slice of tissue about 0.5 cm thick, with the more superficial structures at the top of the screen, and deep structures at the bottom.

The time taken for the crystal to create a pulse of sound is approximately one millionth of a second, and the crystal then 'listens' for about one thousandth of a second. Hence during scanning ultrasound is within the body for only one thousandth of a second for every second of scanning time.

Ultrasound passes easily through water, with very little sound reflected back to the transducer, so that all water- or fluid-containing structures appear black on the ultrasound picture. The full bladder, therefore, acts as a 'window' into the pelvis (Fig. 6.1); and similarly amniotic fluid allows us to visualize the fetus. Ultrasound does not pass through air, so in order to

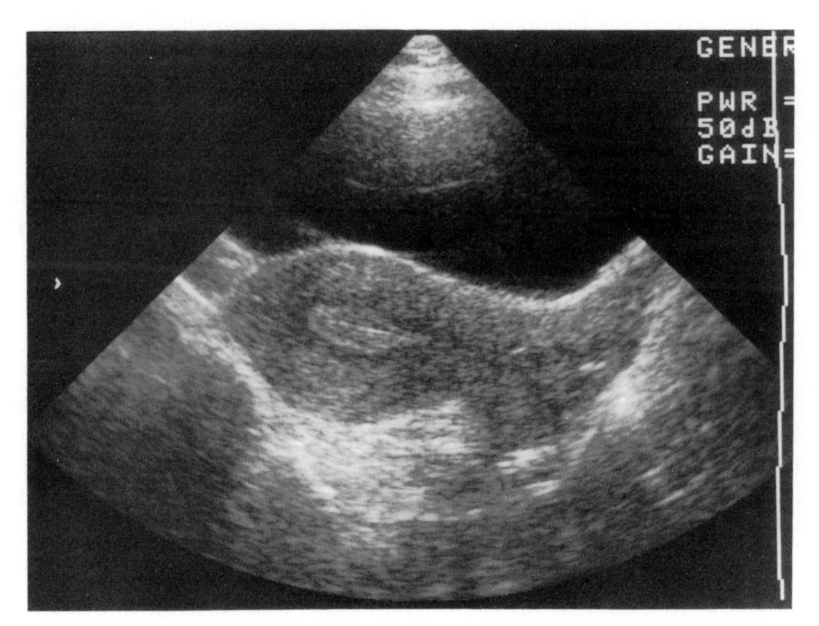

Fig. 6.1a. Normal uterus, longitudinal scan.

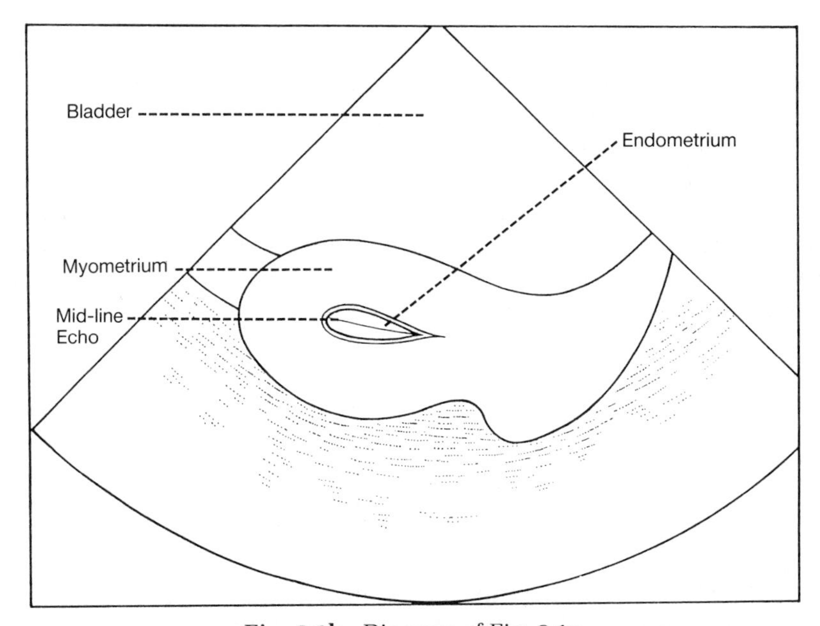

Fig. 6.1b. Diagram of Fig. 6.1a.

scan the abdomen externally a coupling medium—ultrasound gel—is required between the transducer and the skin.

Ultrasound is almost totally reflected back from bony structures, which appear as very dense images. This results in a paucity of echoes behind a dense structure, or 'acoustic shadowing' (Fig. 6.2).

Different tissues have slightly different densities on an ultrasound image: myometrium has a uniform grey texture, whereas placenta is brighter, with a uniform whitish density (Fig. 6.3).

The transducer

There are three basic types of transducer: a linear array, which produces a square image; a sector scanner, producing a sector image; and a curvilinear transducer, producing an image which is a combination of the two. Both first-trimester scanning and gynaecological scanning are usually carried out using a sector transducer, because it is easier to manipulate for pelvic structures. Scanning in the second and third trimester is carried out using a linear or curvilinear transducer.

The machine

There are a number of controls on the machine which can be used to alter the image. The most important are the GAIN controls. The gain control amplifies the signal returning sound-waves produce on the television screen. As a rule, sound-waves reflected from deep structures need more amplification than those reflected from superficial structures.

The gain controls are usually divided into NEAR GAIN, FAR GAIN, and OVERALL GAIN. Increasing the near gain will make the most superficial structures (upper part of the image) brighter. Similarly, increasing the far gain makes the deeper structures (lower part of the image) brighter. The overall gain control will brighten the image as a whole. Some machines have a bank of slider controls to adjust the gain at different levels, which probably gives more precise control.

Some machines have an AUTO button, which will automatically amplify deep structures to produce a constant, well-balanced image. Even if there is no automatic control, once the machine is set up to produce a good image, it is only in very obese and very thin patients that the gain will have to be altered. For obese patients, the gain will need to be increased, and for very thin patients, decreased.

The POWER output of the machine is the amount of energy entering the tissues, and is usually preset. On some machines this may be varied, and by increasing the power one will increase the brightness of the picture.

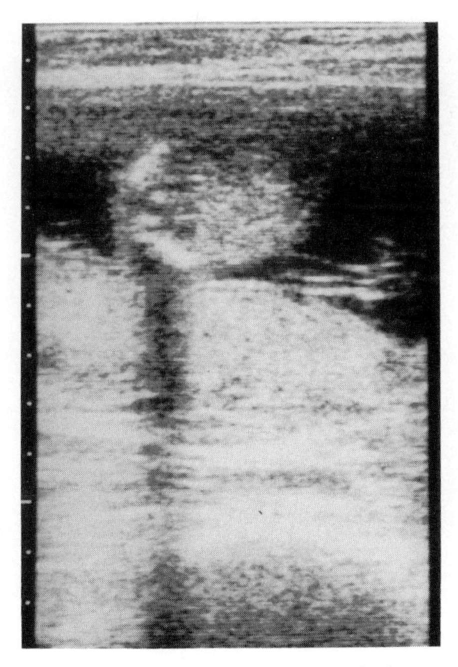

Fig. 6.2a. A transverse scan through a 16-week fetus, showing acoustic shadowing from the fetal pelvis.

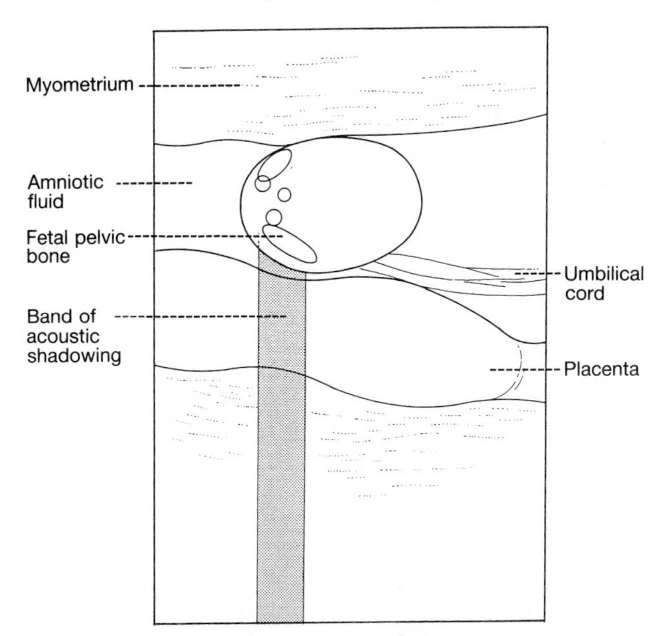

Fig. 6.2b. Diagram of Fig. 6.2a.

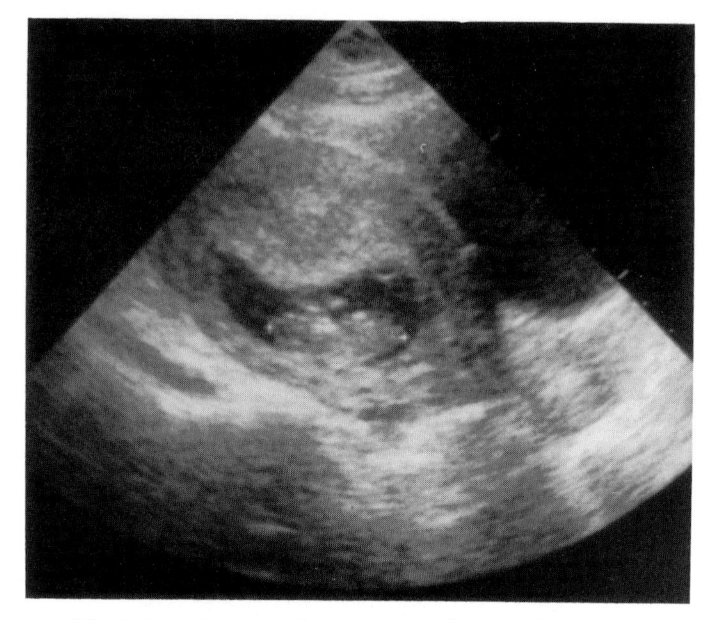

Fig. 6.3a. A ten-week pregnancy, longitudinal scan.

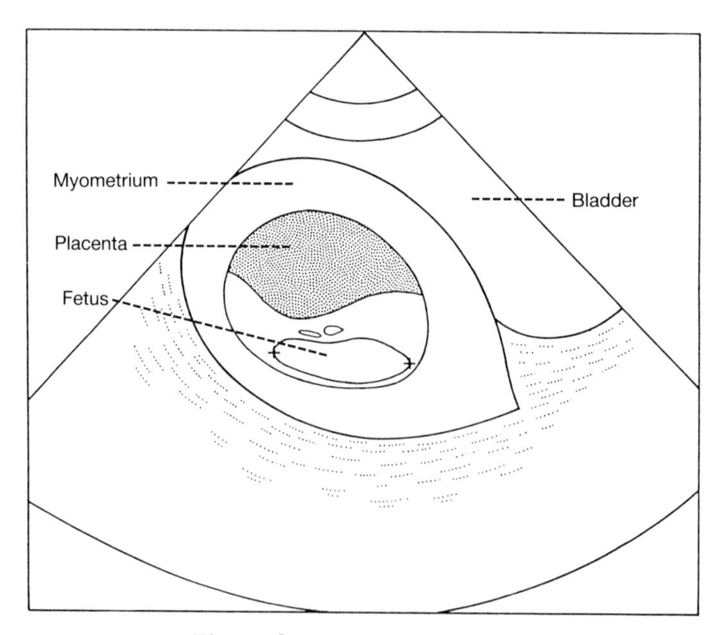

Fig. 6.3b. Diagram of Fig. 6.3a.

However, it is desirable to keep the power output as low as possible, and to use the gain controls to adjust the image.

Images can be magnified, and direct measurements can also be made from the screen using electronic calipers.

Technique

The most important point of technique for abdominal scanning during early pregnancy is a full bladder. After 10–12 weeks' gestation, the uterus is large enough to be seen on scan without a full bladder. The advent of vaginal scanning, however, may well change this situation over the next few years.

Patients are requested to attend for scanning with as full a bladder as possible. In practice this is difficult to achieve, so most patients spend time 'filling up'. If the bladder proves empty on an initial scan patients are given two litres of water or orange juice to drink, and then rescanned after 30–40 minutes.

For scanning the patient is placed supine, and warm gel is applied to the lower abdomen. The person performing the scan sits on the patient's right-hand side facing the patient. Scans are carried out in the longitudinal and transverse planes.

By convention, when scanning transversely, the left side of the image corresponds to the right side of the patient (as if looking at the patient's abdomen from the foot of the bed). In longitudinal scans, the left side of the image points to the head, and the right side to the feet.

Before scanning it is important to know which end of the transducer represents the left side of the image. This can easily be ascertained by sliding a finger with some gel over one end of the transducer, and a shadow will form on screen corresponding to that end of the image. Orientation can then be established as described above.

Oblique scans can also be carried out, but in practice it is better to start with transverse and longitudinal scans to orientate oneself. With experience, one finds that one is constantly changing the position of the transducer subconsciously in order to obtain the clearest image. Scanning movements should be made slowly, so that one can follow the outline of structures and see how they relate to one another.

6.3 Normal anatomy

A longitudinal scan through the pelvis will demonstrate the uterus lying behind the bladder. The uterus has a characteristic pear shape, and the myometrium has a homogeneous grey texture. Within it is a white mid-line echo, representing the apposed walls of the uterine cavity (Fig. 6.1). The cervix is visible as the rounded lower border of the uterus. The vagina extends from the cervix along the inferior border of the bladder. The vagina and uterus rest at an angle of about 80–90° to one another when the bladder is full.

The endometrium appears as a low-density band on either side of the white mid-line echo, and increases in size during the first half of the menstrual cycle. At the time of ovulation, the endometrium becomes dense, and produces an 'ovulation ring' (Fig. 6.4).

Following ovulation, the endometrial echoes become denser, and, if the patient becomes pregnant, this endometrial reaction continues as the decidual reaction. This decidual reaction is not specific for an intra-uterine pregnancy, but may also be seen with an ectopic pregnancy, or if the patient's period is delayed for some other reason. Similar appearances may be seen with an incomplete abortion or with retained products of conception.

The ovaries are small oval structures on either side of the uterus. They have a density lower (darker) than that of the uterus, and, depending on the stage of the menstrual cycle, will contain variably sized follicles, which appear as round black areas with the ovaries (Fig. 6.5).

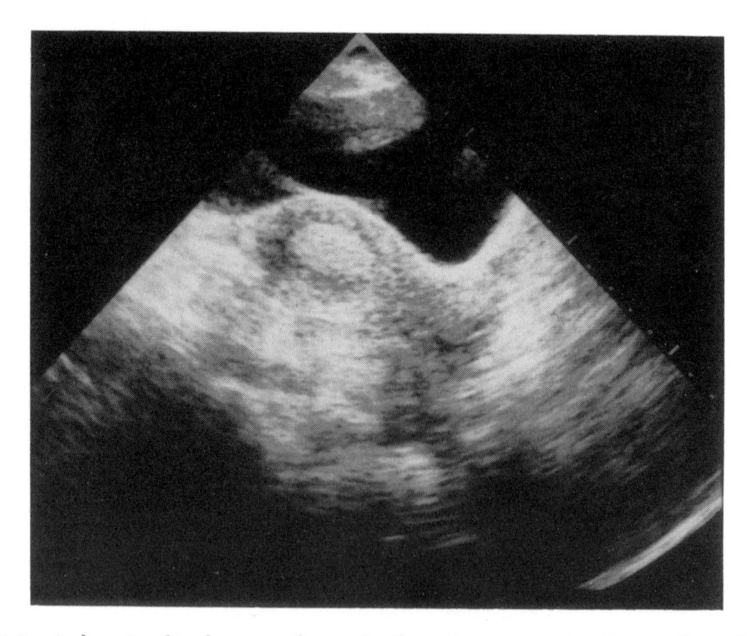

Fig. 6.4. A longitudinal scan through the uterus at the time of ovulation, showing a dense endometrium.

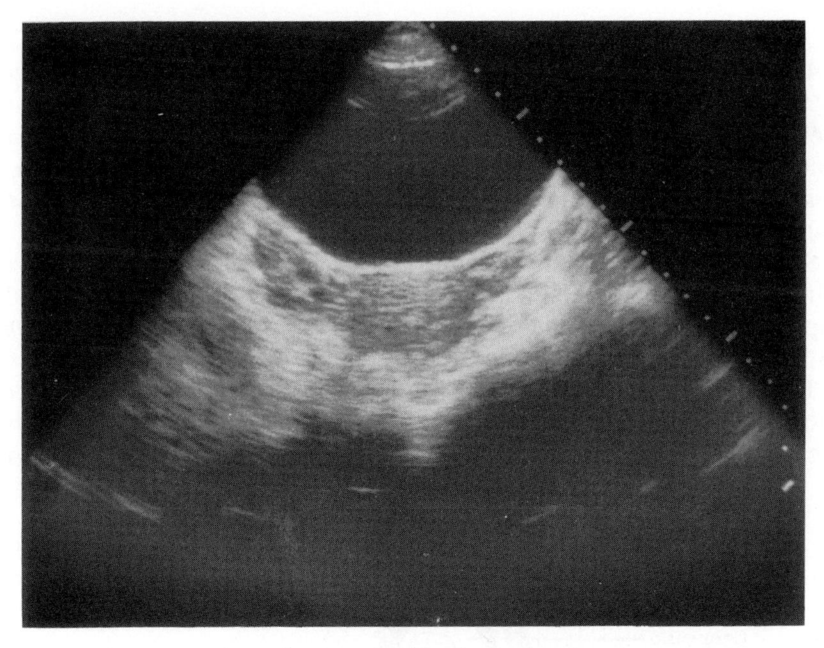

Fig. 6.5a. Normal uterus and both ovaries, transverse scan.

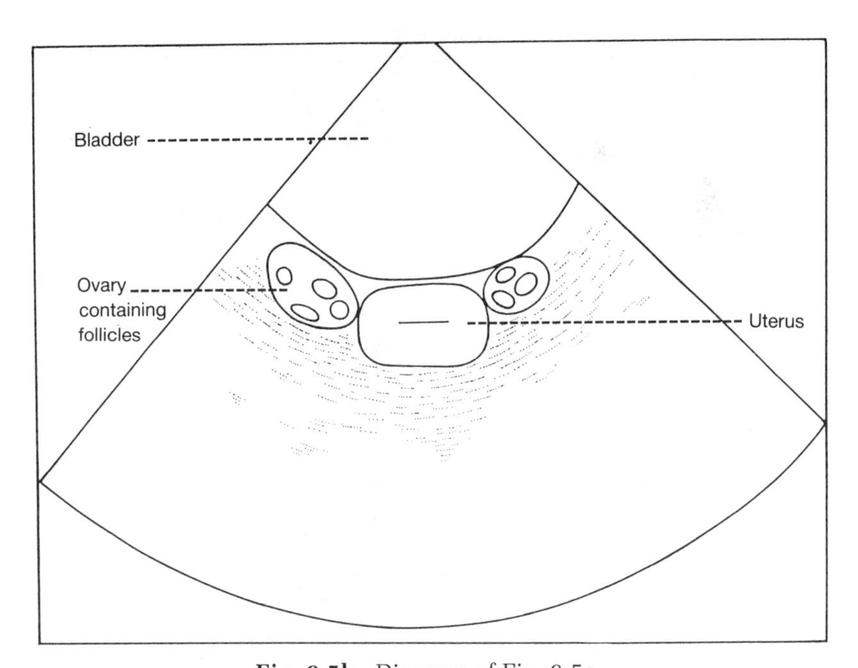

Fig. 6.5b. Diagram of Fig. 6.5a.

6.4 Embryology and ultrasound appearances

Gestation age is usually calculated from the time of ovulation for the embryologist, and from the first day of the last period for obstetrician. In this discussion, all gestational ages will be from the first day of the patient's last menses.

Soon after implantation within the uterus, the decidual reaction differentiates into the decidua basalis at the site of the implantation, the decidua capsularis which surrounds the gestation sac, and the decidua parietalis over the remaining uterine wall (Figs 6.6 and 6.7).

The gestation sac is seen as a well-defined black circle, usually eccentrically placed within the uterus. The sac is made up of the embryo, within the fluid-filled amniotic cavity, and also the fluid which lies between the

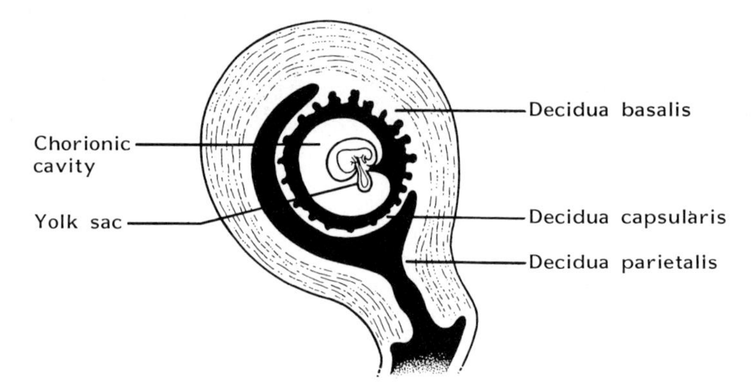

Fig. 6.6. Diagrammatic representation of an eight-week pregnancy.

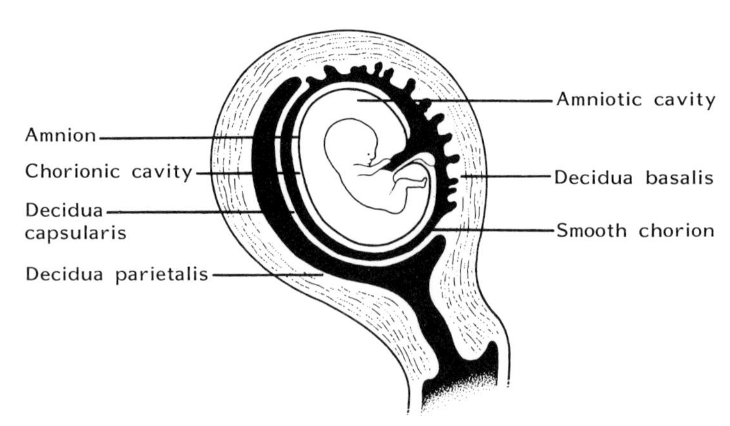

Fig. 6.7. Diagrammatic representation of a twelve-week pregnancy.

amniotic cavity and the chorionic cavity, the extra-embryonic coelom. By 9–10 weeks' gestation the amniotic cavity completely fills the chorionic cavity, with apposition of the amnion and chorion. Occasionally, the amniotic membrane is seen as a parallel line within the gestation sac.

The decidual reaction is seen around the gestation sac as a dense white ring. Frequently a double ring will be seen, representing the two decidua, capsularis and parietalis (Figs 6.8 and 6.12, p. 000). The decidua basalis, which later forms the placenta, may be present as a thickening of part of the inner ring, producing some asymmetry. The 'double ring' sign may be helpful in differentiating an intra-uterine pregnancy from the decidual cast seen associated with an ectopic pregnancy.

The mean gestation-sac diameter is a useful measurement, and is calculated by measuring three diameters of the sac, usually anteroposterior, transverse, and longitudinal, and then dividing by three. The diameter of the gestation sac is approximately 5 mm at four weeks' gestation (Fig. 6.9), and increases by 1.0–1.5 mm per day for the next four weeks (Fig. 6.10).

A viable pregnancy cannot be established until the fetal heart movement is seen; this is usually present from six to seven weeks of gestation. The fetal heart may well be seen before the fetal pole is visible, and is visualized as a constant flicking motion, at a rate of about 120–140 beats per minute, within the sac.

The fetal pole is present from around seven weeks, and appears as a well-defined density within the gestation sac (Fig. 6.11). It can be difficult to demonstrate the fetal pole if it lies close to the wall of the sac; hence the value of demonstrating the fetal heart flicker.

At 7–8 weeks gestation, one can often see the yolk sac as a small, well-defined white ring within the gestation sac (Fig. 6.12). It usually measures 3–5 mm in diameter.

The fetal pole can be measured. Measurement is taken from one end of the fetus to the other. This measurement is known as the 'crown–rump length', and there are standard graphs of crown–rump lengths versus gestational age. This measurement is accurate to within ± 5 days in the first trimester (Fig. 6.13). Care must be taken in early pregnancy not to include the yolk sac in the measurement (Fig. 6.11). Later in pregnancy the fetus tends to flex, and this will produce smaller measurements, so making one liable to underestimate the size of the fetus. After 14 weeks the crown–-rump length measurement is no longer used, and the biparietal diameter measurement can be used to assess gestational age.

From eight to twelve weeks' gestation the fetus is readily defined, and individual body parts are discernible. The placenta becomes evident as a distinct structure from about eight weeks onwards, and appears as an eccentrically placed dense white band along one edge of the gestation sac (Fig. 6.3, p. 000).

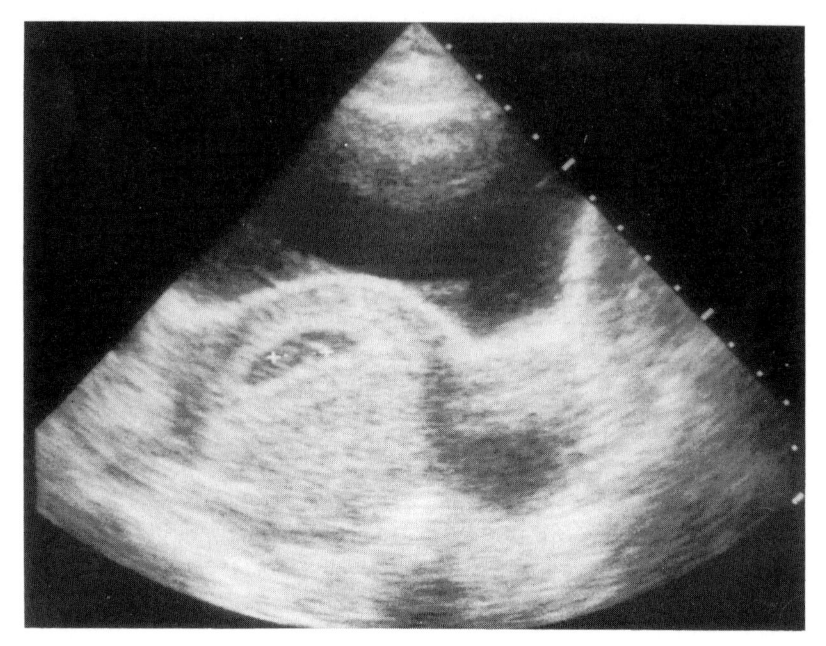

Fig. 6.8a. Six-week pregnancy and corpus luteum cyst, transverse scan.

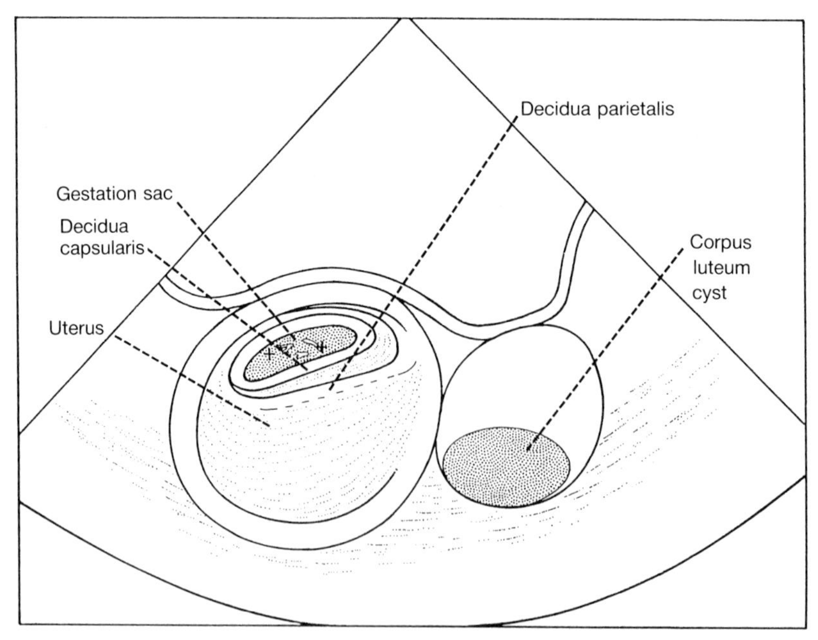

Fig. 6.8b. Diagram of Fig. 6.8a.

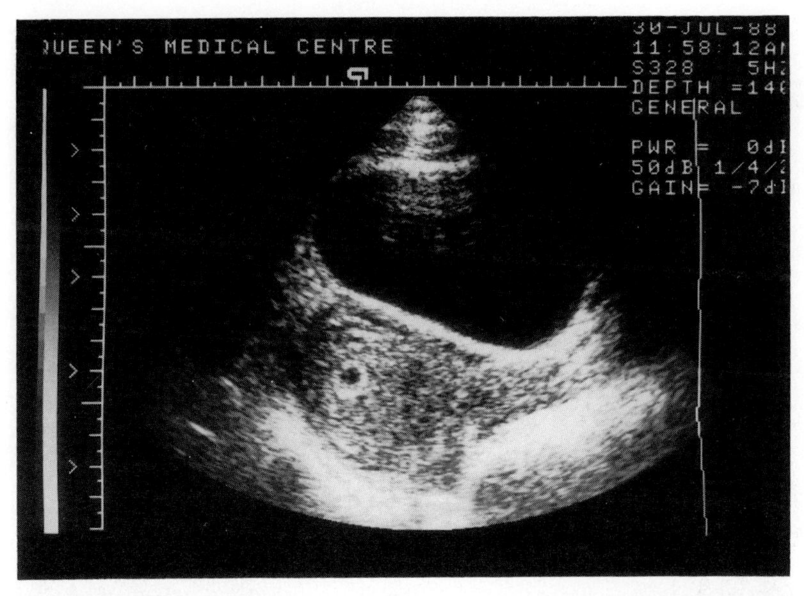

Fig. 6.9. Four-week pregnancy, showing a 0.5 cm gestation sac, longitudinal scan.

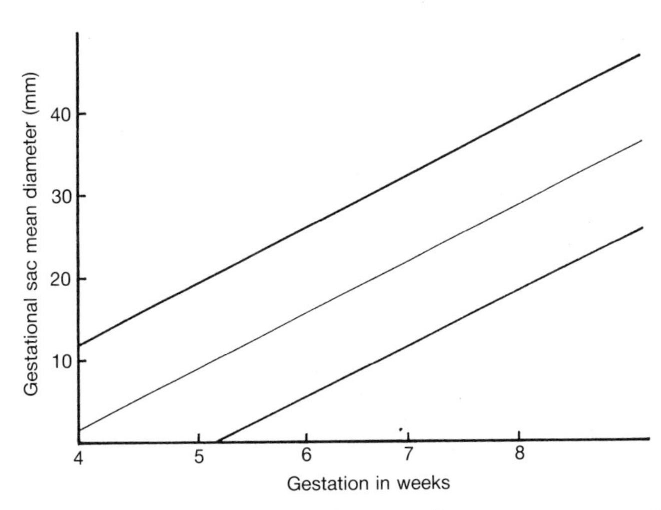

Fig. 6.10. Graph showing mean gestation sac diameter versus weeks of gestation, reproduced with permission from Batzer *et al.* (1988).

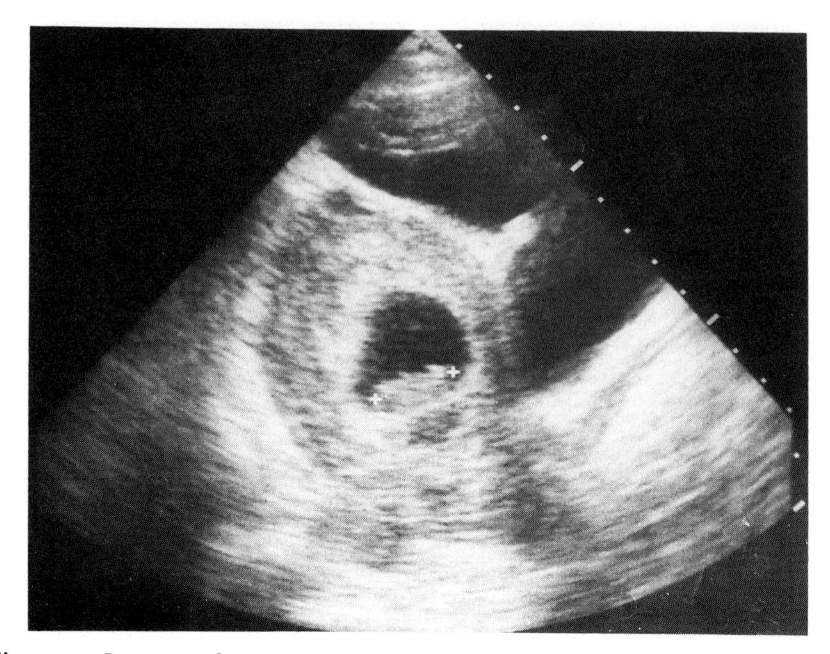

Fig. 6.11. Seven-week pregnancy, showing crown–rump measurement of the fetal pole, longitudinal scan.

Weekly change

5 weeks' gestation

The gestation sac meaures 5–12 mm in diameter, and surrounding it is a dense decidual reaction appearing as a white ring. No fetal pole is visible within the sac at this stage (Fig. 6.14).

6 weeks' gestation

The gestation sac is now 18 mm in diameter, and a fetal heart flicker may be detected. The diagnosis of viability rests on the presence of a fetal heart flicker.

7 weeks' gestation

The gestation sac measures 24 mm, and a fetal pole and the fetal heart should be detected (Fig. 6.11).

8 weeks' gestation

The gestation sac measures at least 3.0 cm, and a yolk sac may be visible. There are a definite visible fetal pole and fetal heart (Fig. 6.12).

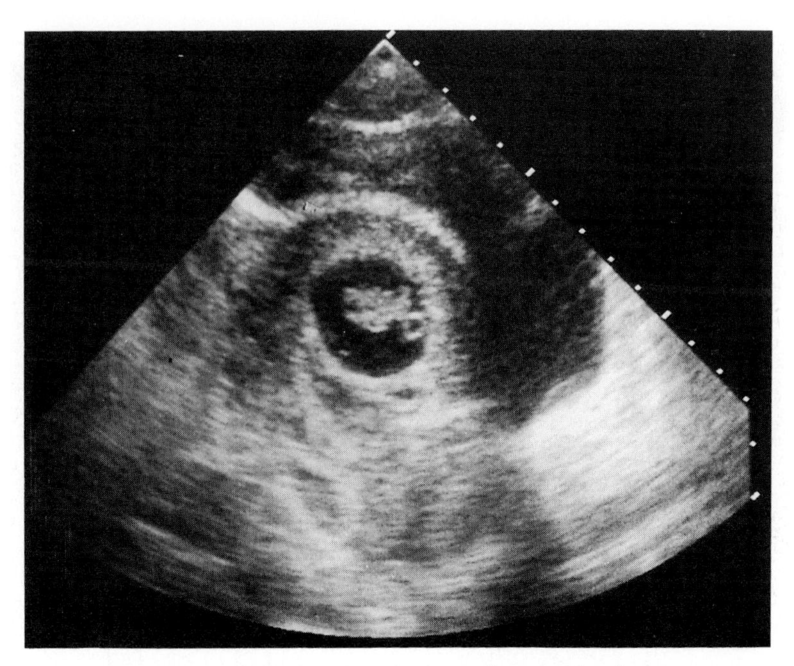

Fig. 6.12a. Eight-week pregnancy, longitudinal scan.

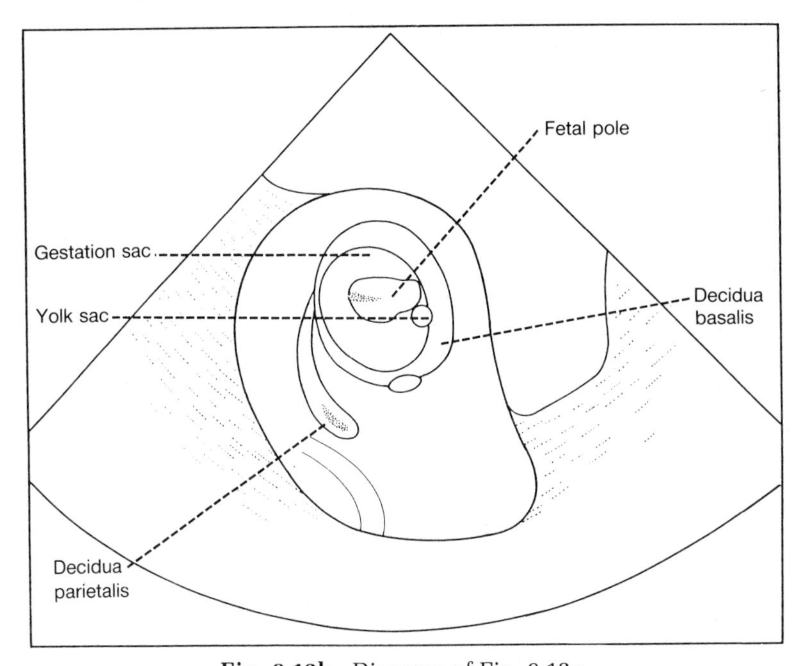

Fig. 6.12b. Diagram of Fig. 6.12a.

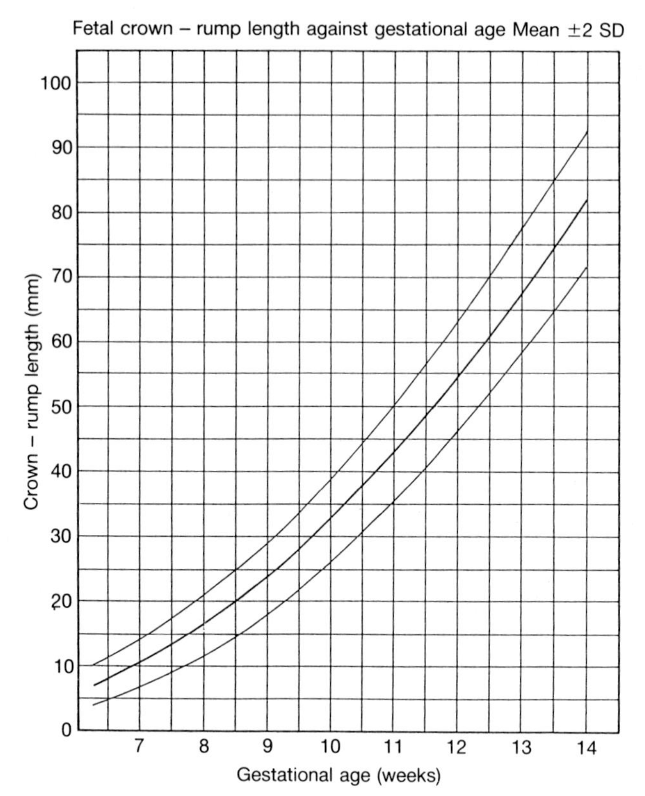

Fig. 6.13. Graph showing crown–rump length versus weeks of gestation, reproduced with permission from Robinson and Fleming (1982).

9–12 weeks' gestation

The anatomy of the fetus is clearly seen. The head, trunk, and limbs are easily visualized (Fig. 6.15).

Twin pregnancy

Diagnosis of a twin pregnancy depends upon the demonstration of two fetal poles. In the case of diamniotic twins a septum will be seen separating the two fetuses (Fig. 6.16). In monoamniotic twins no septum will be demonstrated, and there will be a single placenta.

6.5 Abnormal findings in early pregnancy

Ultrasound scanning is extremely useful for assessing complications of

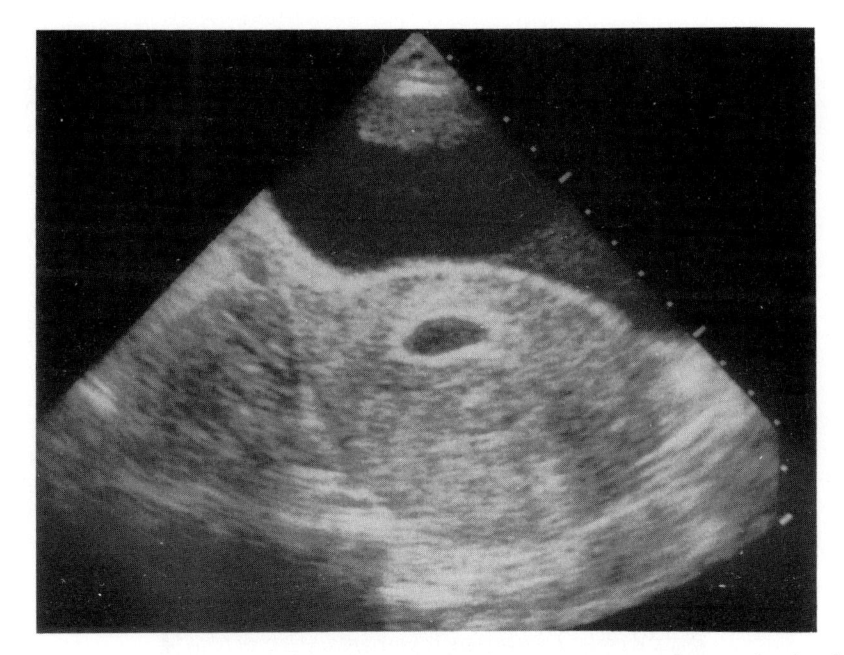

Fig. 6.14. Five-week pregnancy, showing a 1 cm gestation sac but no fetal pole, transverse scan.

early pregnancy, and for demonstrating pathology within the pelvis. Bleeding and pain are common presentations in threatened abortion, so that demonstration of a normal viable pregnancy by ultrasound is very reassuring, as over 90 per cent will progress to term.

In the case of a complete abortion, the uterus will appear essentially normal, with no gestation sac and no decidual reaction. The uterine cavity will be empty, with a central mid-line echo.

The appearances of an incomplete abortion will vary, depending on whether a gestation sac is present or not. If a gestation sac is present, it will be smaller than expected, and will have an irregular wall with no fetal pole (Fig. 6.17); or else the position of the sac may be abnormally low within the uterus. The decidual ring around the sac may be irregular.

The size of the sac is important, because if the sac is smaller than 3 cm in diameter then it could represent a 5–7 week gestation, where the fetal pole has not yet developed. In that situation it is best to give the patient 'the benefit of the doubt', and rescan in 10–14 days to reassess the situation. When the sac is greater than 3 cm in diameter, a fetal pole should be present; but its absence, together with an irregular wall, is good evidence of an incomplete abortion.

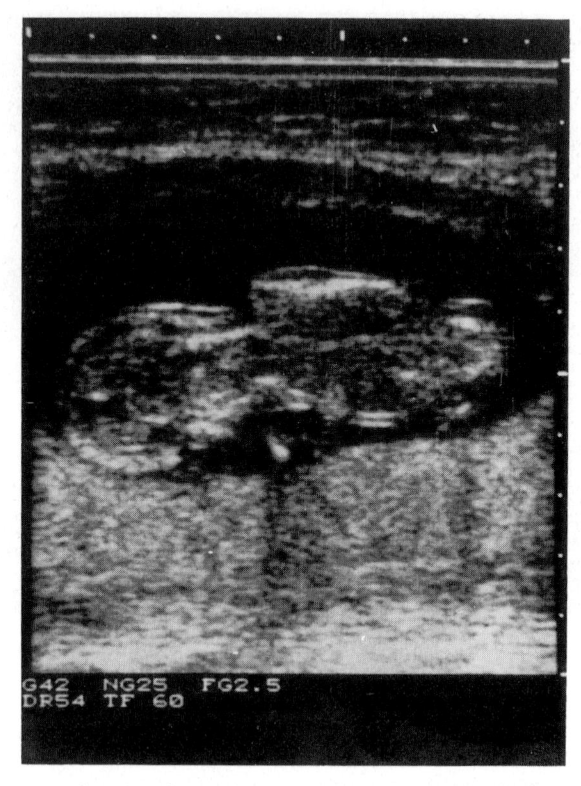

Fig. 6.15. A coronal scan through a fourteen-week fetus, showing head and thorax.

Where the gestation sac has already been lost the uterus contains only decidual remnants and blood clots, and ultrasonically these appear as a dense whitish echo within the uterine cavity (Fig. 6.18).

In missed abortion there is a gestation sac with a fetal pole, but no fetal heart movement is seen. The sac and fetal pole are usually smaller than expected, and the same may have an irregular wall (Fig. 6.19).

Diagnosis of a blighted ovum or anembryonic pregnancy depends on demonstration of a gestation sac greater than 3 cm in diameter without a fetal pole (Fig. 6.20).

Ectopic pregnancy can pose a diagnostic problem; however, the absence of an intra-uterine pregnancy in a patient with a positive pregnancy test is highly suggestive of the diagnosis. There are, in addition, a number of important signs to help confirm the diagnosis. A solid adnexal mass may be seen, which is separate from the ovaries. There may be fluid in the pouch of Douglas (seen as a black area behind the uterus), representing intra-peritoneal blood. A decidual cast may also be seen within the uterus (Fig.

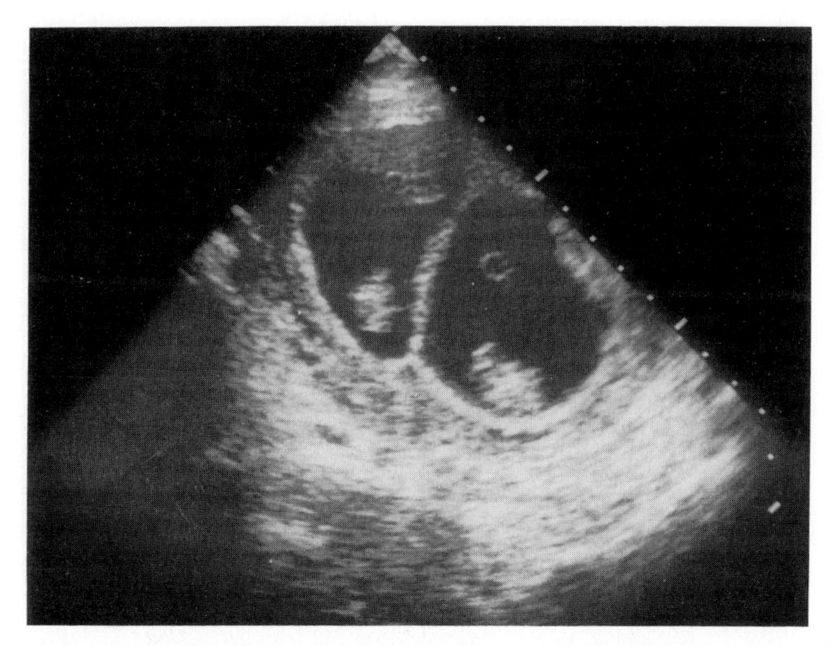

Fig. 6.16. Ten-week twin pregnancy, showing two sacs, two fetal poles, and a yolk sac in one of the sacs.

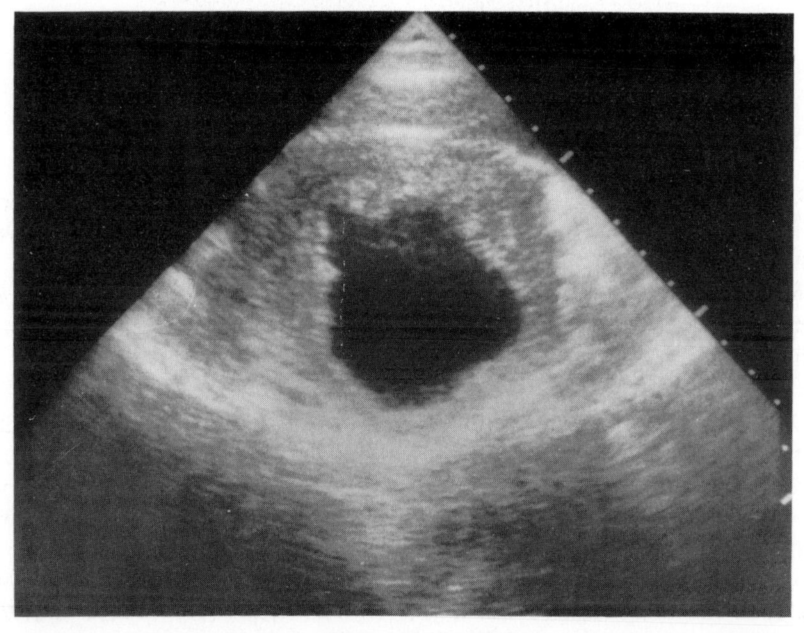

Fig. 6.17. Incomplete abortion, showing a gestation sac with an irregular wall and no fetal pole.

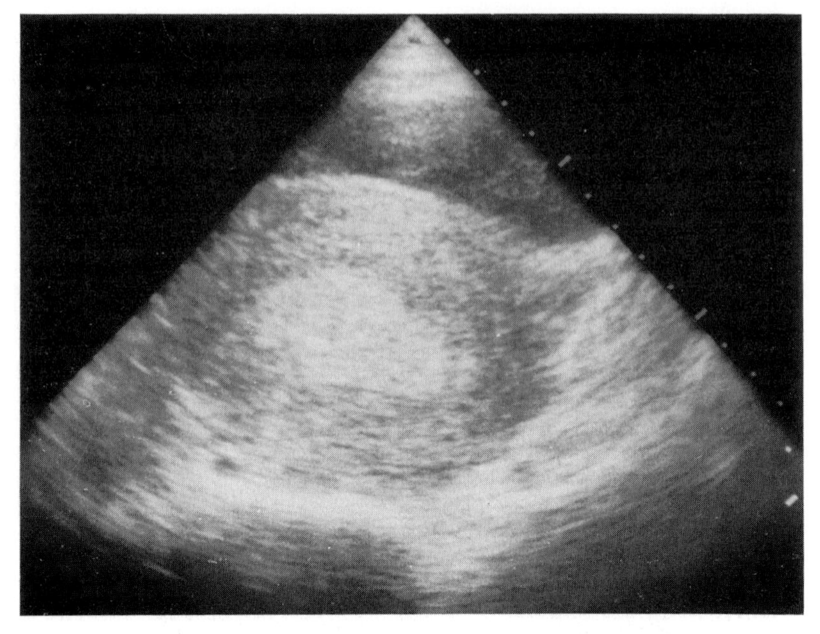

Fig. 6.18. Incomplete abortion, showing only dense echoes within the uterine cavity.

6.21). Occasionally a fetal pole is seen within an adnexal mass which may demonstrate a fetal heart echo (Fig. 6.22).

A molar pregnancy has a characteristic ultrasound appearance. In the uterus is a large dense mass, with numerous dark cystic areas (Fig. 6.23). A high human chorionic gonadotrophin level will confirm the diagnosis.

Fibroids within the uterus are usually well-defined rounded or oval masses within the myometrium, sometimes distorting the mid-line echo. They can be hypoechoic (dark), hyperechoic (white), or the same density as myometrium (Fig. 6.24).

Large ovarian cysts are easily diagnosed, since the ovary is replaced by a well-defined black circle which, if simple, is uniloculated (Fig. 6.25).

Finally, venous sinuses or lacunae can often be seen in first-trimester placentae. In sampling these areas are best avoided, since sampling within them can cause bleeding (see Fig. 8.1, p. 118).

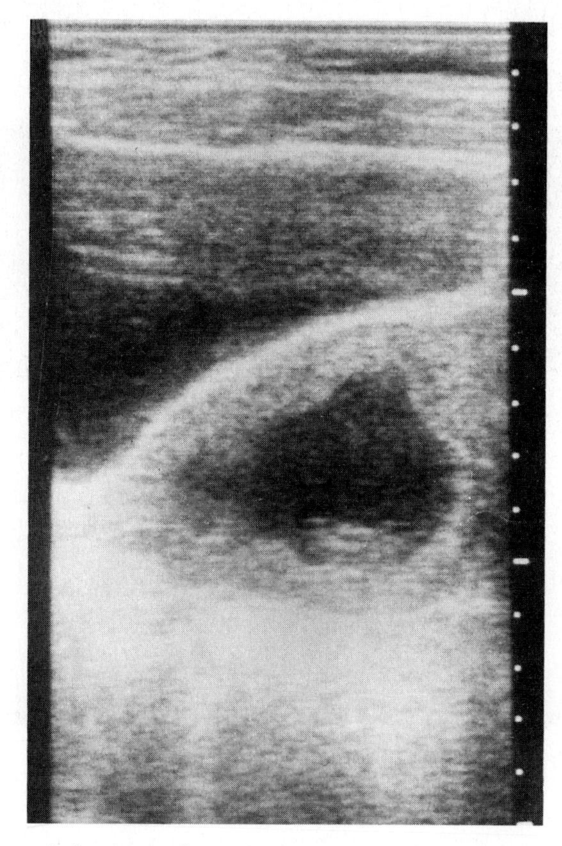

Fig. 6.19. Missed abortion, showing a gestation sac with an irregular wall; no fetal heart movement was detected within the fetal pole.

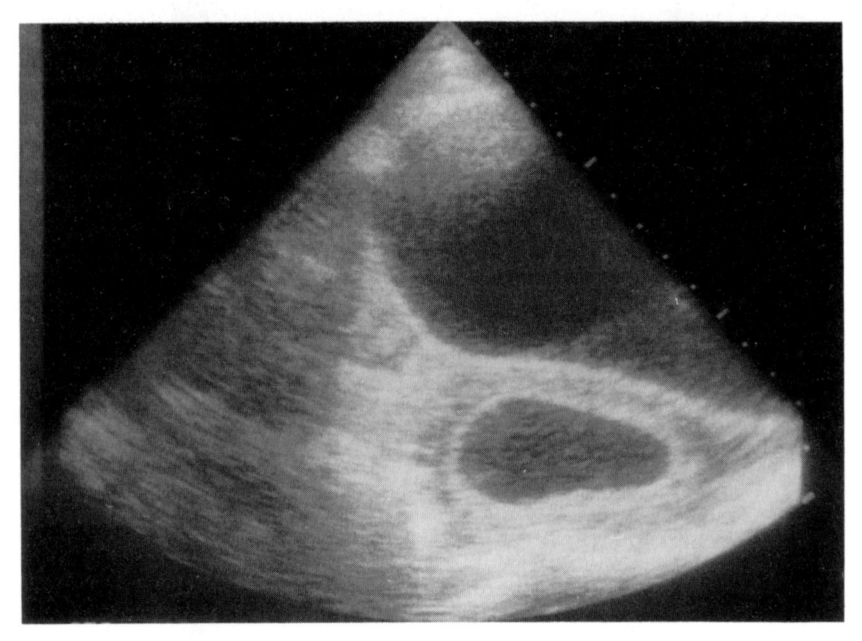

Fig. 6.20. Blighted ovum: a smooth gestation sac greater than 3 cm diameter with no fetal pole.

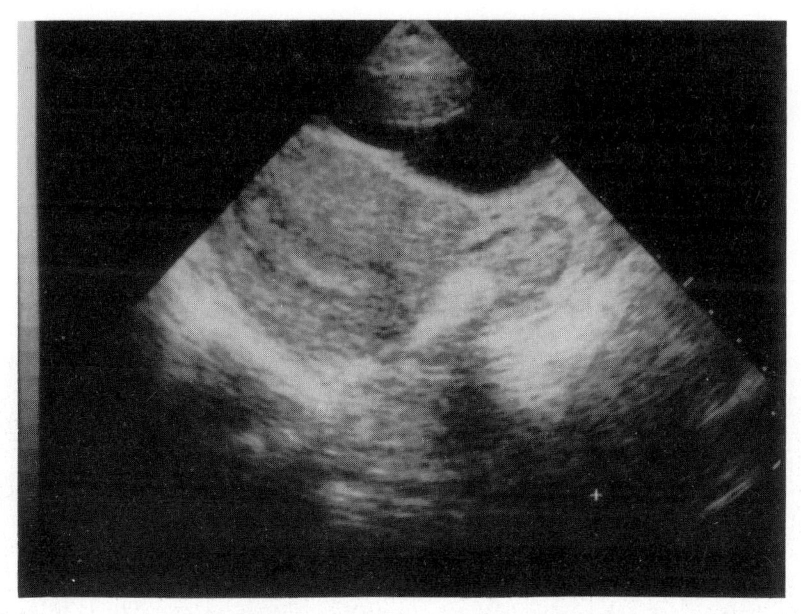

Fig. 6.21a. Ectopic pregnancy: there is a solid adnexal mass with a dense decidual reaction within the uterus.

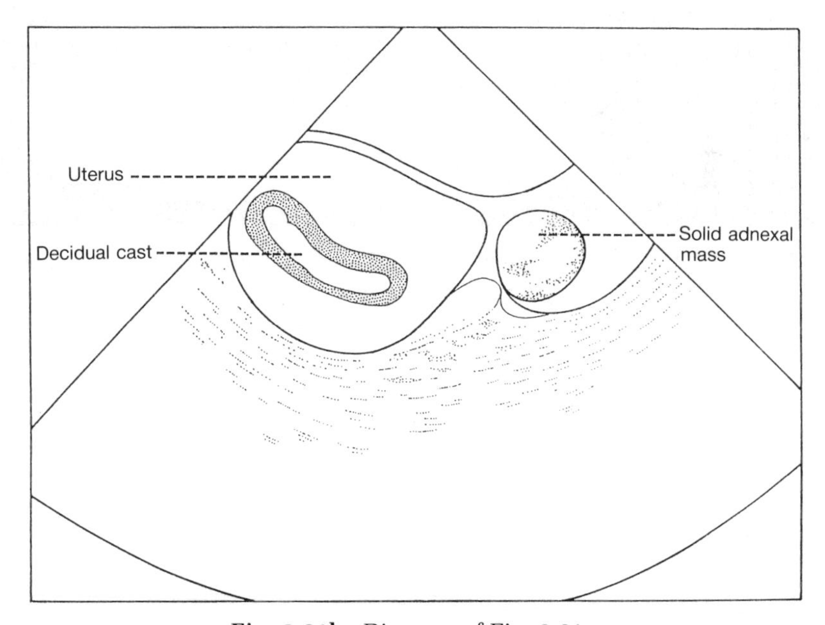

Fig. 6.21b. Diagram of Fig. 6.21a.

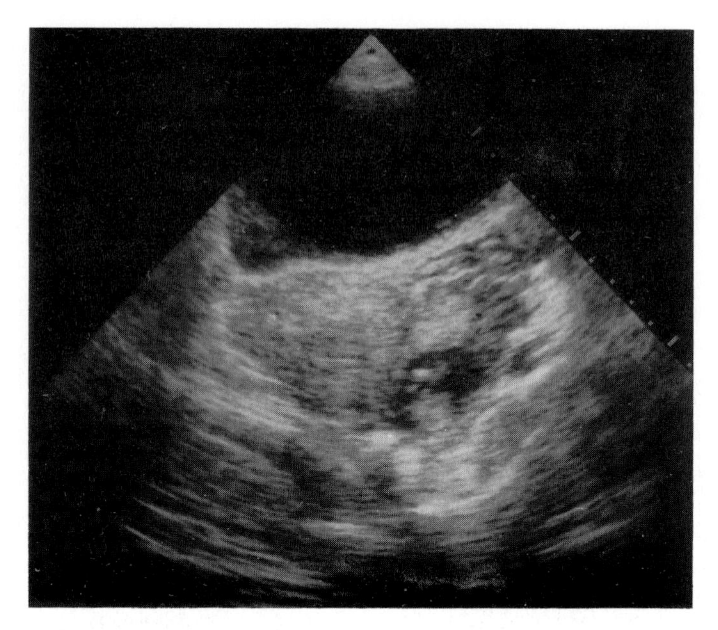

Fig. 6.22a. Ectopic pregnancy: there is a cystic adnexal mass containing a fetal pole. Fetal heart movements could be detected within the fetal pole.

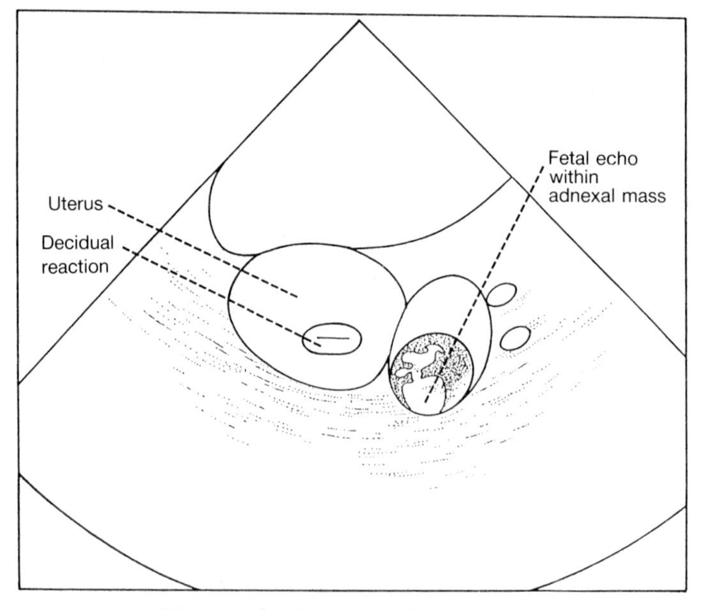

Fig. 6.22b. Diagram of Fig. 6.22a.

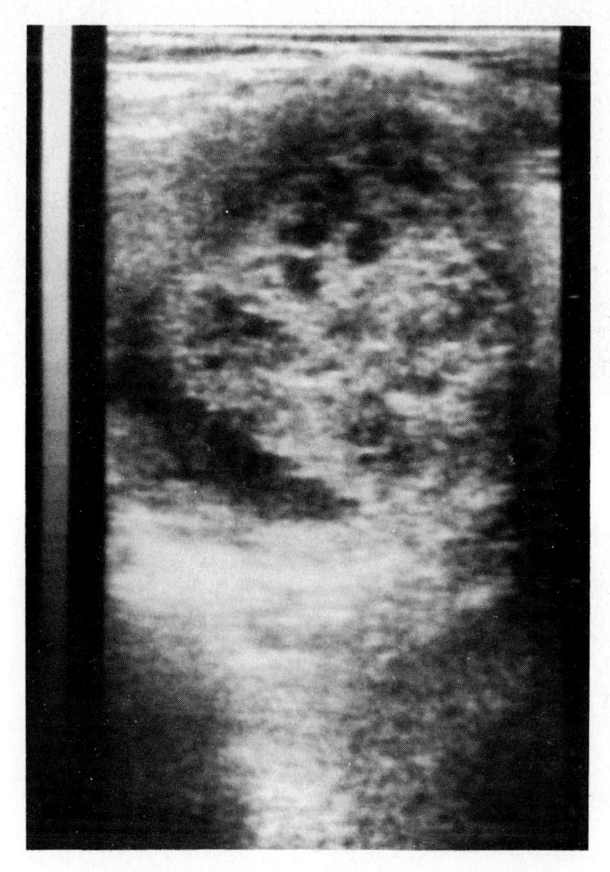

Fig. 6.23. Hydatidiform mole: the uterus contains a solid mass with multiple cystic spaces.

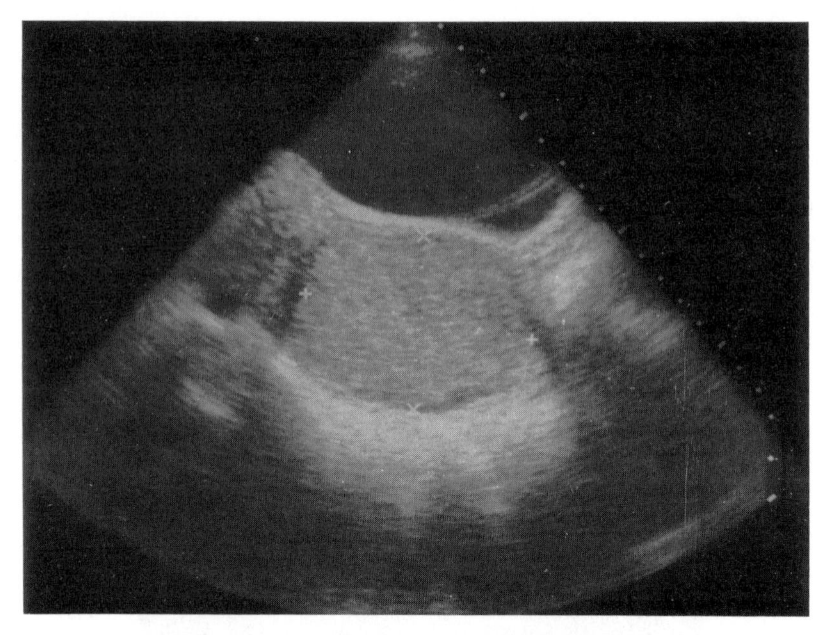

Fig. 6.24. Fibroid: there is a well-defined oval area of increased density within the uterus.

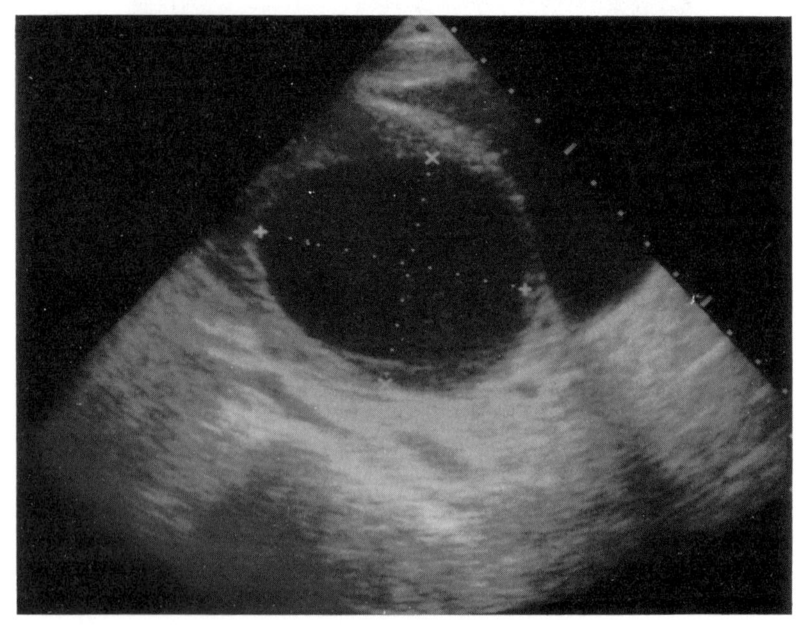

Fig. 6.25. Ovarian cyst: a well-defined echo-free area within the ovary.

References

Batzer, F. R., Weiner, S. W., Corson, S. L., Schlaft, S., and Otis, C. (1988). Landmarks during the first 42 days of gestation. *American Journal of Obstetrics and Gynaecology*, **146**, 977.

Robinson, M. P. and Fleming, J. E. E. (1982). A critical evaluation of sonar crown–rump length measurements. *British Journal of Obstetrics and Gynaecology*, **82**, 702–10.

7 Ultrasound and chorion villus sampling

Rosalyn E. Richardson

7.1 Introduction

Ultrasound is one of the tools available to the obstetrician performing chorion villus sampling. With ultrasound scanning the contents of the uterus can be visualized, thus enabling optimal positioning of the sampling cannula. There is less danger to the fetus, because the movement of the cannula can be monitored, and the chances of obtaining an adequate sample the first time are improved. Ultrasound is used throughout pregnancy to monitor the growth of the fetus and to keep a check on its well-being. The majority of obstetricians are very familiar with the images obtained.

An ultrasound beam travelling through a complex medium is modified whenever a change of acoustic properties is encountered. Either scattering or reflection occurs, with the result that some ultrasound is transmitted back to the probe. The returned sound echoes are used to form an image of the medium under interrogation. The size of the reflections depends on the variations in the acoustic properties of the medium. In the body the acoustic properties of the different tissue types are all approximately the same, with the notable exception of bone. This means that when an ultrasound beam encounters a change in acoustic properties the size of the echo that is returned to the probe is small, and most of the beam travels on through the body. When performing chorion villus sampling under ultrasound control the operator introduces into the uterus a sampling device, usually a fine cannula or needle, whose properties are very different from those of human tissue. This can complicate the procedure, since artefacts are produced which can obscure the image. The artefacts vary according to the type of machine being used, so to fully understand the origin of the artefacts and how to avoid them it is necessary to understand how the ultrasound picture is produced. This chapter will give a basic description of the different types of ultrasound machine available, with their characteristic problems and advantages when the transcervical approach is used.

7.2 Chorion villus sampling under ultrasound control

The uterus should be scanned so that a comprehensive picture of the contents is obtained. Once the best site for sampling has been located the probe should be manipulated so that the sampling site, cervix, and vagina are visualized together. This shows the obstetrician the path that the cannula should follow to give the best chance of obtaining a successful biopsy. When the cannula is inserted it is important to be able to see along its entire length, otherwise it is impossible to be certain that the end of the cannula seen on the screen is the true end, and not simply the point at which the cannula moves out of the scan plane. Incorrect location of the end of the cannula is an important cause of unsuccessful biopsies. To obtain the required view may mean that the probe must be held at quite an angle to the skin. Maintaining this position is harder with some probe designs than with others, particularly as the skin will be covered with a coupling medium which will tend to make it slippery.

7.3 Ultrasound machine types

Modern ultrasound equipment gives a real-time image of the part of the body under investigation. The screen picture is made by rapidly sweeping an ultrasound beam over the site of interest. The speed of ultrasound through the body is such that a complete picture is composed at a rate which is too fast for the eye to detect the movement of the sound-beam. The composite picture appears as a film image on the screen.

There are two main methods of generating real-time ultrasound images. The first method uses a single ultrasound crystal mounted so that it can oscillate through an angle of 90°. Alternatively, many crystals can be used, and activated sequentially. The first method is termed a mechanically steered sector scanner, usually simple referred to as a 'sector scanner'; the second, an array scanner.

Sector scanners

Sector-scanner probes produce a fan-shaped image. The ultrasound beam is steered through an angle of 90° either by rocking the crystal back and forth or by placing the crystal in front of an acoustic mirror that oscillates. The speed of oscillation is such that the crystal has not significantly altered position before all echoes of interest have returned from the body. Ultrasound travels sufficiently fast through the body that this can occur at relatively rapid oscillation rates. The advantage of this system is that the motion of the crystal is not apparent to the observer. In addition, movement within the body, such as the beating of the fetal heart, is visualized. The sector angle is limited to 90° because, although the centre of the image is renewed with each oscillation, the edges are only refreshed every other frame. For wider angles the frame rate would not be sufficiently rapid to prevent the loss of apparent real-time imaging at the edges of the picture.

Sector probes are usually light and easy to manoeuvre. The mechanical parts, however, tend to wear out, and the probe may need replacing before an array probe purchased at the same time; but they are much cheaper than array probes. Focusing is at a fixed depth, achieved either by using a lens attached to the crystal or by shaping the crystal. Electronic focusing is not possible with such probes, unless they are so called annular probes. These probes are made up of several concentric rings of crystal, and focusing is achieved by phasing the firing of the probe segments. This type of probe is not available on all machines.

For chorion villus sampling the sector probe has several advantages. Because it produces a fan image it is possible to obtain a view underneath obstructions such as the pubic bone, or bowel gas that is anterior to the uterus. The small size of the probe face means that there are no contact problems.

The probe does however also have some disadvantages. The design of the probe means that the scan plane is not immediately apparent from the probe position. In addition the sector-scan image has a small anterior field of view, and this can cause problems where the site of interest is near to the skin surface.

Another type of sector-scanner probe contains three or four separate crystals mounted on a circular former that can rotate. These probes have the advantage that wider angles are possible, because the image is always refreshed from the same side. Focusing is fixed using mechanical means, as discussed above. The probes are far more bulky than single crystal probes, and this may result in positioning problems, though the scan plane is easier to identify. A sector probe of this type is shown in Fig. 7.1a, with a typical image in Fig. 7.1b.

Array scanners

There are two main types of array scanners, linear arrays and phased arrays. Array probes contain many small ultrasound crystals: a typical linear array probe will have about 200 crystals, and a phased array of the order of 100. The crystals are activated in groups, in order to produce a beam that has sufficient power to penetrate the body and is non-divergent except in the far field. Electronic focusing is used with these probes, which means that the focal zone can be adjusted. This is an advantage for chorion villus sampling, as the best sites for sampling are not always at the same depth in the body.

Linear arrays: Six to eight crystals are fired together to produce a beam of ultrasound. Once echoes from the deepest structure of interest have

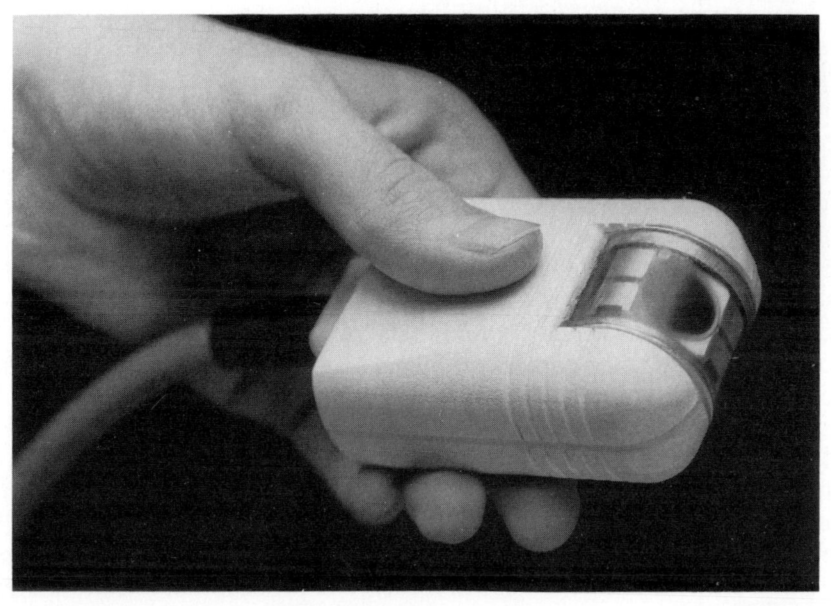

Fig. 7.1a. Sector probe.

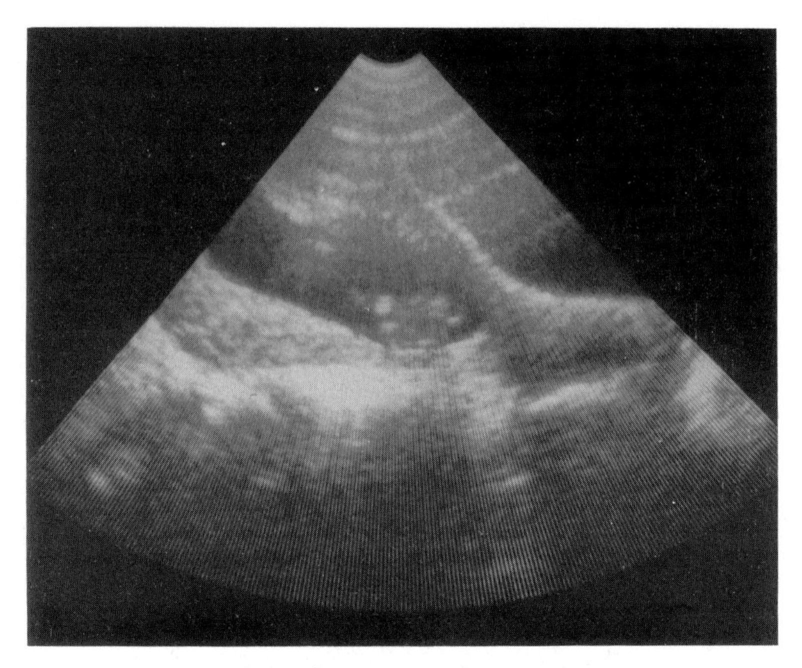

Fig. 7.1b. Sector-probe image.

returned to the probe the next group of crystals is activated. To ensure that the lines of ultrasound are close together the active group of crystals moves along the array a crystal at a time. Linear array probes tend to be quite large—typically a 3.5 MHz probe is about 150 mm long—and this can cause contact problems for some probe orientations. If the sampling site is low in the pelvis the pubic bone may block the view with a linear array probe, whereas a probe with a small probe face could improve visibility. The advantage of the larger probe is that it is very easy to identify the scan plane. A linear array probe is shown in Fig. 7.2a; Fig. 7.2b shows a typical image.

Linear arrays are also available with a curved probe face, which produces a sector image. These probes are larger than either single-crystal sector probes or phased arrays, but combine the advantages of both. The curved probe face causes fewer contact problems than the straight linear array; but the dimensions are such that the scan plane can be easily identified. A curvilinear probe and the associated image are shown in Fig. 7.3.

Phased arrays: Phased array probes are much smaller than linear array probes. The crystals are activated in groups like a linear array, but are not fired together. Instead a time-delay is introduced between firings, to direct the beam at an angle. The exact angle is determined by the phasing of the crystal activation. The advantage of this type of probe is that a sector image is obtained without having any moving parts. The angle of a phased array

Fig. 7.2a. Linear array probe.

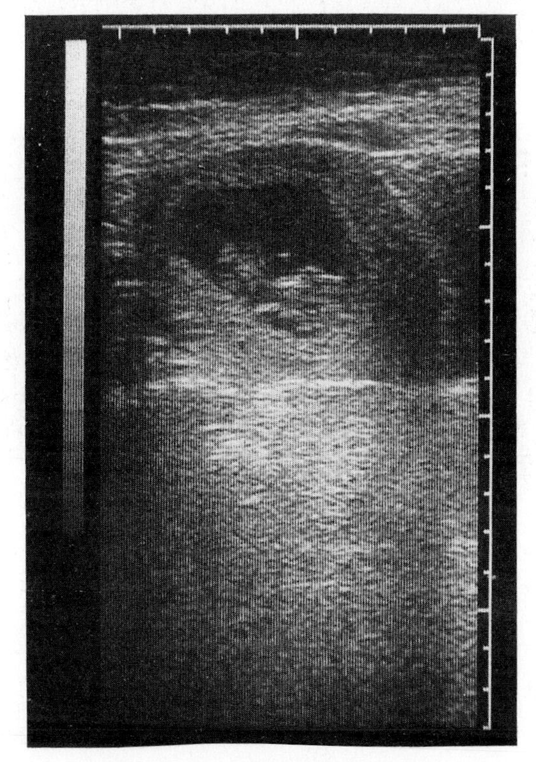

Fig. 7.2b. Linear array image.

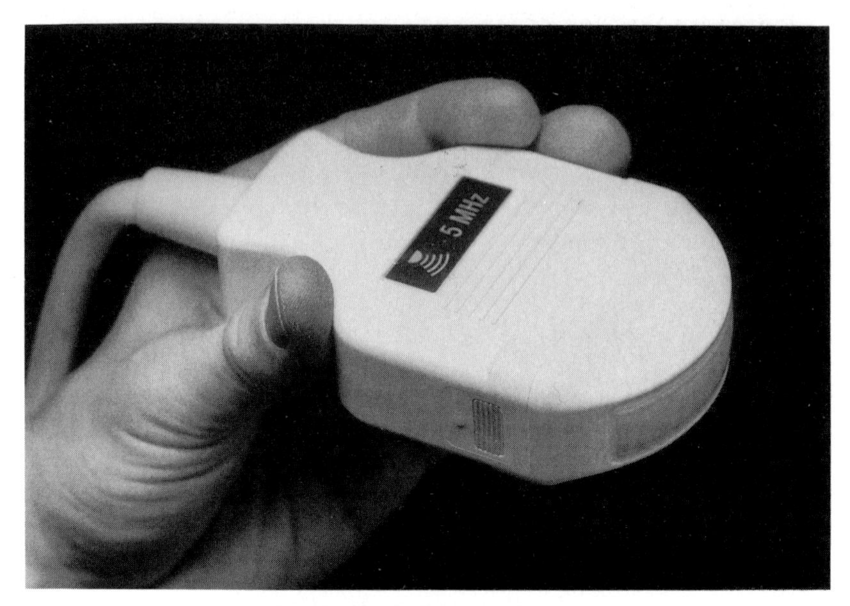

Fig. 7.3a. Curvilinear probe.

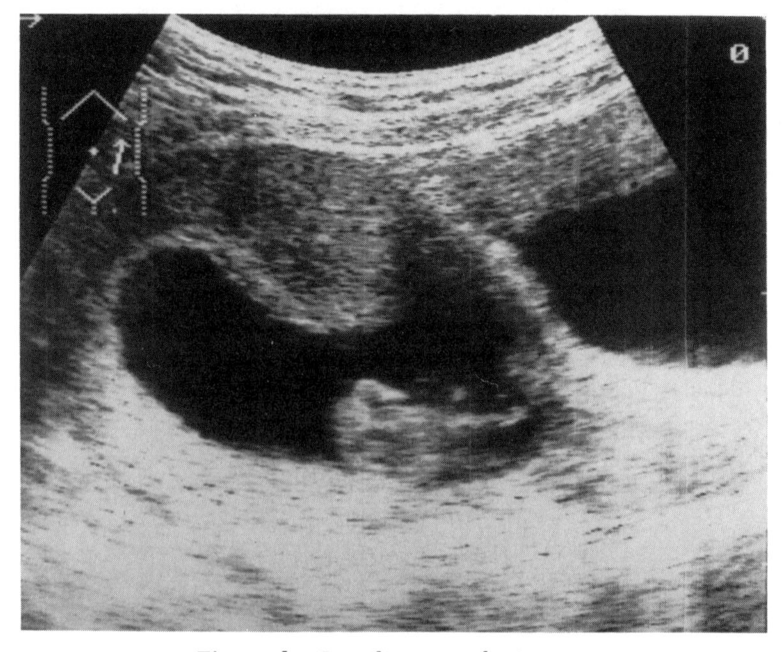

Fig. 7.3b. Curvilinear probe image.

image is limited to 90°, because there are problems with directing the beam at acute angles. Since an array of crystals is used not only can the beam be directed at different angles, but it can also be focused electronically. The probes have a rectangular face about 25 mm long, and although the scan plane is not so easily identified as with a linear array the probes are very light and easy to use. The small size does mean that there are no contact problems, and the sector image can improve visibility. Fig. 7.4 shows a phased array probe and image.

Linear arrays are excellent at giving a complete picture of the uterus at the stage of pregnancy when chorion villus sampling is carried out. The dimensions of the probe mean that the scan plane is easily identified, but may also cause contact problems. Sector and phased array probes are much more compact, and do not suffer from contact problems. The sector image may not contain a complete section of the uterus; but this may not be a disadvantage. Bowel gas may obscure the uterus when scanned with a linear array; and an improved image can sometimes by obtained with a sector scan. Identification of the scan plane is easiest with a linear array probe.

7.4 Ultrasound image artefacts

There are many well known artefacts of ultrasound that affect obstetric images. It is not the intention of this section to discuss these artefacts in

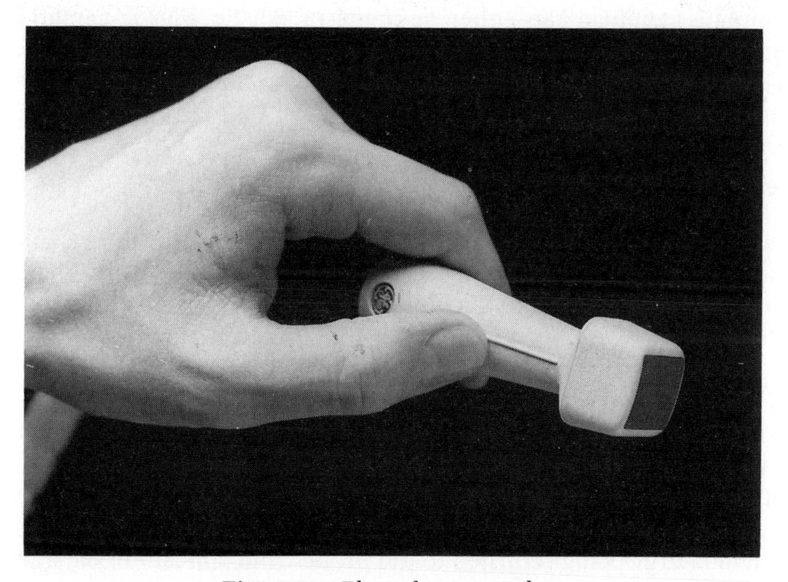

Fig. 7.4a. Phased array probe.

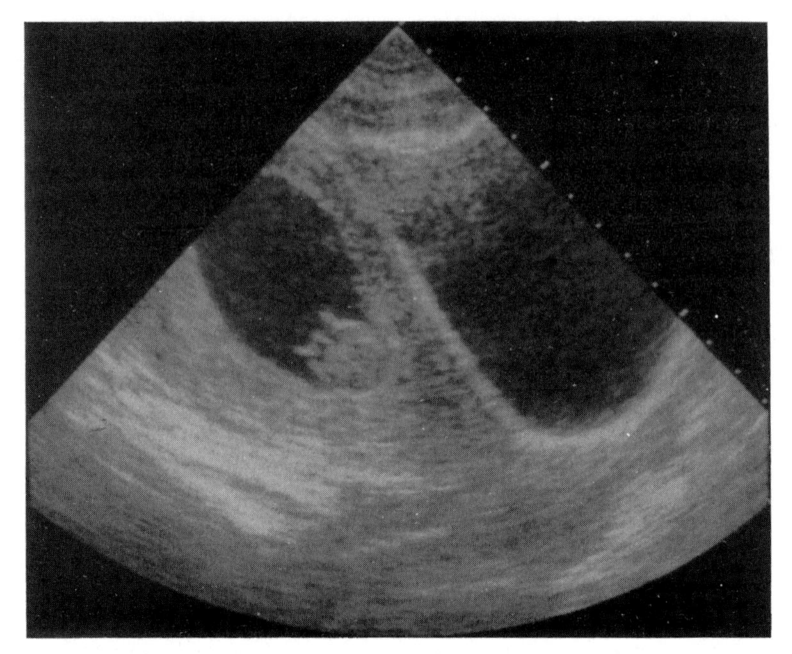

Fig. 7.4b. Phased array image.

detail: they have been fully described elsewhere. The artefacts under discussion are those peculiar to sampling procedures.

Sampling cannulae can either be metal or plastic. Plastic cannulae are harder to see on an ultrasound scan, but do not cause as many artefactual images as metal cannulae. The majority of the artefacts from imaging metal cannulae are related to the beam width.

Slice-width artefact

Ultrasound images consists of a thin slice through the body; and the thickness of the slice is governed by the focusing of the beam. The cross-section of the beam perpendicular to the scan plane is wider than the effective beam width. The beam width is narrower than the true cross-section of the beam, because although there is energy outside the beam, echoes returned to the probe from interfaces in this region are below the threshold of detection of diagnostic ultrasound machines. This is true as long as the reflection coefficient is of the order of the typical values of body tissue. Metal cannulae have very different properties to body tissue, and reflect a much higher proportion of incident energy. This results in sufficient energy being returned to the probe from a metal–tissue interface in the tail of the beam to register as a valid echo. The effective beam width is therefore much wider for strong reflectors such as metal. The result of this on the image is that the cannula appears to be where it is not. A typical image is shown in Fig. 7.5. The cannula is actually between the sac and the wall of the uterus; but in this image it appears to be right through the sac. This artefact is more readily seen with array probes, where the beam can only be focused in the scan plane. This results in a beam with non-circular symmetry, and the dimensions of the beam perpendicular to the scan plane can be considerably greater than the dimensions in the scan plane. This artefact cannot be avoided; but the operator must be able to identify it as erroneous.

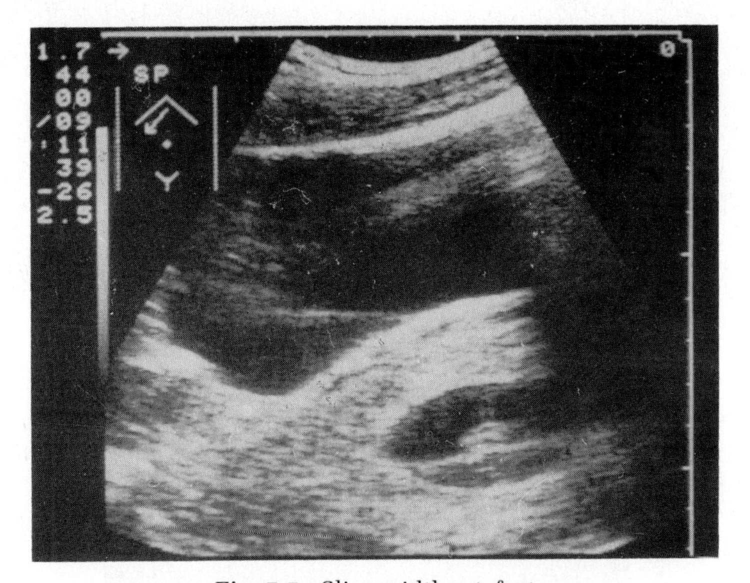

Fig. 7.5. Slice-width artefact.

This artefact is easy to distinguish from images where the cannula is tenting or infringing on to the membranes. When the sac is disturbed in this way the contents will move, and this movement can be seen on the real-time image. In addition the sac wall often shows a sharp indent where the cannula is impinging into the membranes. Tenting of the membranes is illustrated in Fig. 7.6 below.

Beam-width artefact

As with the slice-width artefact, the effective beam width in the scan plane is greater for strong reflectors than for weak reflectors. This can be seen in ordinary obstetric images, particularly when the bladder is full. The posterior bladder wall returns a very strong echo to the probe, because the returning signal is amplified, as if the beam were attenuated during its passage through the bladder. If the uterus indents the bladder an artefact can be generated from the anterior wall of the uterus.

The metal Liu sampler has a solid end, and can be positioned to demonstrate the beam-width artefact. If the chosen sampling site is anterior it may be necessary to manipulate the cannula so that it is end-on to the beam. The solid tip is a very strong reflector, and in this orientation will return echoes to the probe from a much wider area than the tip-size warrants. A typical image is shown in Fig. 7.6. The beam-width artefact is seen with all probe types, and can usually be eliminated by angling the probe.

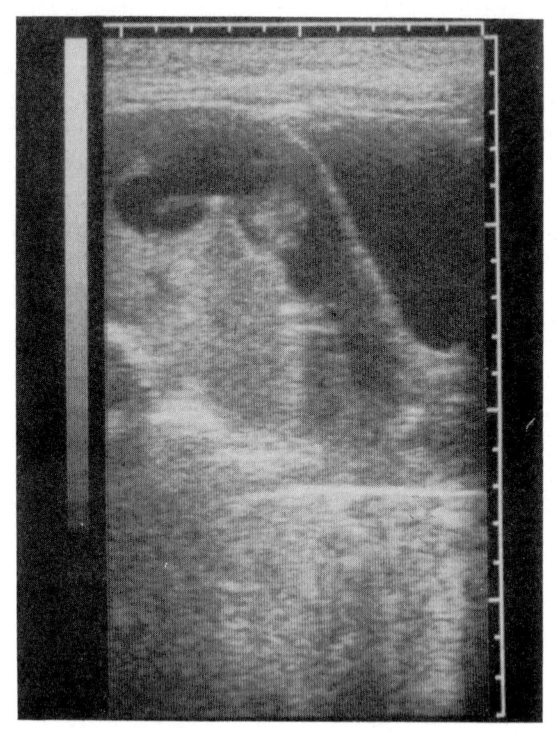

Fig. 7.6. Beam-width artefact from the solid cannula tip.

Diffraction-grating artefact

This artefact only occurs with array probes, and not all array probes demonstrate it.

An array probe consists of many ultrasound crystals separated by a small gap. This arrangement is equivalent to a larger source of ultrasound with a fine grating in front of it. The resulting beam has side-lobes such as are found with light-sources shone through a diffraction grating. The design of array probes purposely minimizes the energy in the side-lobes; but some residual energy is left. This appears on images from array probes like an exaggerated form of the beam-width artefact. In fact it can be hard to distinguish between the two. The artefact becomes particularly apparent when a metal cannula is perpendicular to the beam. This arrangement is most likely to occur with a linear array, where each beam of ultrasound is perpendicular to adjacent beams. The cannula is a strong reflector, and when the cannula is perpendicular to the beam there is sufficient energy in the side-lobes to register at the probe. The result of this on the image is that the cannula appears to be much further into the patient than it actually is. It is difficult in these situations to identify the location of the tip of the cannula. The artefact may be eliminated by increasing the angle of the cannula or the probe. This reduces the amount of side-lobe energy reflected back to the probe.

Figure 7.7a shows the cannula tip apparently located in the section of the uterus obscured by bowel gas. In fact there is no clear end to the cannula: the intensity of the image fades into the bowel-gas region. When the cannula is angled (Fig. 7.7b) the tip is easily seen.

Reverberation

Reverberation is usually characterized by multiple echoes of diminishing intensity displayed at fixed intervals. The sound energy is reflected back and forth between two interfaces, and at each interface some of the energy is transmitted and some reflected. The beam is also absorbed as it travels through the tissue, so the energy remaining in the beam diminishes gradually. This is seen in the decreasing intensity of the reverberant echoes.

When ultrasound reverberates through metal a different picture is shown. This is because metal has two important characteristics as far as ultrasound is concerned:

(1) the speed of sound is much greater than in tissue; and

(2) the beam is not absorbed to the same extent as it is in tissue.

The effect of the speed of sound is that if the dimensions of the metal causing the reverberation are small then the reverberant echoes will be too close together to display as separate echoes. Instead, a streak will be seen on the image. The streak will be relatively intense because of the lack of absorption of the sound within the metal. A typical image is shown in Fig. 7.8. When this artefact is seen it clearly demonstrates the end of the cannula, and is therefore a help rather than a hindrance. The artefact is seen with all probe types if a cannula with a solid end is used.

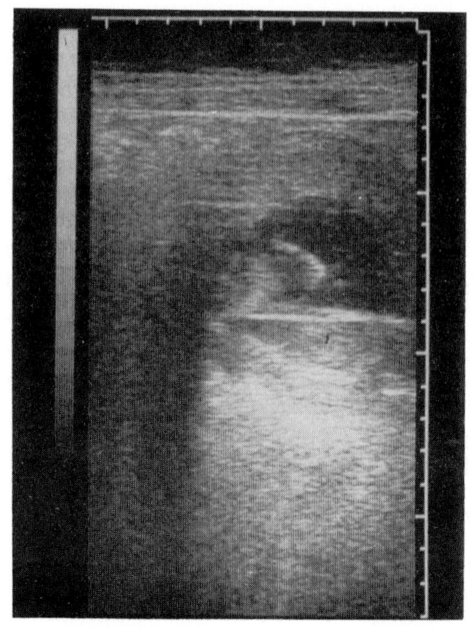

Fig. 7.7a. Diffraction grating artefact demonstrated when the cannula is perpendicular to the ultrasound beam.

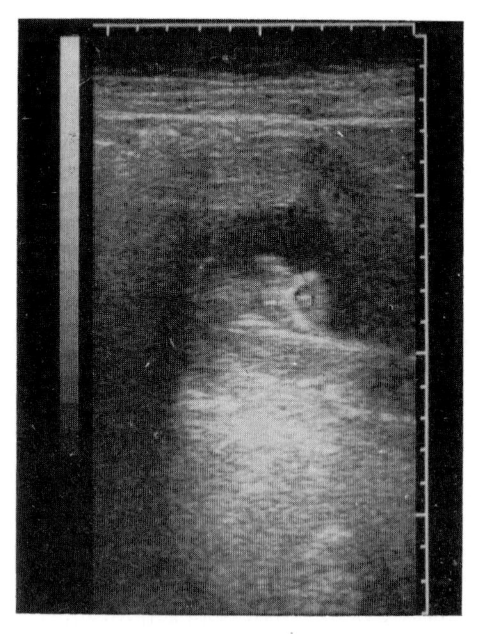

Fig. 7.7b. Angling the cannula removes the artefact from the image.

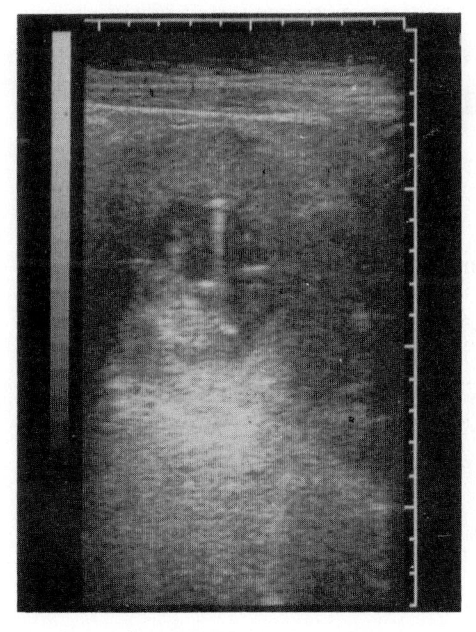

Fig. 7.8. Reverberation from the solid cannula tip.

7.5 Conclusion

Ultrasound is a powerful tool in the hands of an experienced operator. For successful chorion villus sampling it is necessary to understand the possible artefacts and to avoid being misled by them. Interpretation of ultrasound images is not as simple as it appears at first glance.

8 Transcervical chorion villus sampling

David T. Y. Liu

8.1 Introduction

The transcervical approach is currently the most widely used technique for first-trimester chorion villus sampling. Many implements have been proposed as suitable, but only ultrasound-guided collection of villus material with a pair of fine forceps and aspiration through plastic or metal cannulae remain as practical options. Which implement is used rightly remains the obstetrician's prerogative. The principle of the approach is the same. This chapter endeavours to provide sufficient background discussion to allow an informed choice of implement, and to advise the operator through the necessary steps of the technique.

8.2 Pre-operative counselling

For both the transabdominal and transcervical techniques a detailed history from the patient is essential. History-taking allows time for rapport to develop, but more importantly gives an opportunity to explore the reasons behind the referral for villus sampling. Some requests, such as screening for maternal age-related chromosomal abnormality, can be counselled along similar lines as for amniocentesis, while other requests may merely reflect the need for a more precise explanation of the diagnostic potential from villus material. Risks of the more complicated inheritable conditions should be examined in conjunction with support, if it is needed, from clinical geneticists and calculations by molecular biologists. The risks to the fetus and mother undergoing villus sampling should be compared with that for amniocentesis. A woman with a Caesarean section scar may benefit by first-trimester diagnosis, and thus avoid anxieties associated with mid-trimester therapeutic abortion. It is always necessary to assess the whole situation: adopt a holistic attitude, and derive for the couple the safest and most reliable diagnostic option. Understandably there are conflicting anxieties for patients and doctors alike, when on the one hand there is a need to secure a diagnosis, but this must be balanced with the need to minimize the threat to the fetus on the other. This situation can be particularly stressful for the subfertile woman who has achieved a pregnancy through fertilization *in vitro*, only to find a need to exclude chromosomal abnormality. For the woman under forty years of age the option to use maternal serum alpha-fetoprotein or unconjugated oestriol and chorionic gonadotrophin screening in the second trimester as additional pointers before intervention should be discussed (Cuckle *et al.* 1987). Counselling should always be non-directive. Thorough examination of all possibilities for the couple avoids recrimination should mishap arise.

Individual steps of the procedure should be carefully explained. This often reduces anxiety, and fosters the intent to make the patient as relaxed as possible under the circumstances.

Partners are encouraged to be present throughout the procedure. They often appreciate the reassurance that undue discomfort is not caused and that the procedure does not harm the fetus. Apart from their supportive role, partners usually find the experience enhances bonding if all is well, and are more sympathetic if results are adverse.

8.3 Pre-operative preparations

The lithotomy position is used for transcervical villus sampling. The patient selects a position which she finds comfortable, and, if she wishes, one where she can obtain a view of the scan picture. Since abdominal scanning is required, she is advised to relax the rectus abdominis muscles. Contraction of these muscles can lift the scanning probe some 3–4 cm, making it difficult to maintain a consistent picture.

No analgesia or premedication is required for transcervical chorion villus sampling, which can readily be accommodated on an out-patient basis. On the very rare occasion when the highly anxious patient cannot relax sufficiently for sampling to be undertaken with safety, there is no reason why general anaesthesia on a day-case basis cannot be used. Most patients do not find transcervical villus sampling uncomfortable, particularly when the procedure is refined to avoid the use of instruments such as teneculums. The author's own study (Liu *et al.* 1988) showed that patients rated the degree of pain associated with sampling as less than 3 on a linear scale ranging from 1 to 10. Insertion of the speculum was the most uncomfortable step. Using the same method of assessment for pain, transcervical villus sampling compares very favourably with amniocentesis. Furthermore, the same study indicated that a relaxed open atmosphere can reduce the level of anxiety, and is thus more conducive to patients' acceptance of the diagnostic procedure.

A full bladder is advised for abdominal scanning in the first trimester. This is not always necessary for villus sampling, but is dependent on the build of the patient, the presence of bowel gases, and the position of the uterus. A slim patient with an anteverted uterus is allowed or asked to empty a full bladder to minimize discomfort. A full bladder provides a clearer view when the patient is obese, and is essential if the uterus is retroverted. The full bladder can also be used to push away loops of bowel encroaching on the uterus.

The risk of introducing infection by a transcervical approach has caused initial concern (Garden *et al.* 1985; McFayden *et al.* 1985; Wass and Bennett 1985). This anxiety is not surprising, since many asymptomatic woman harbour commensals or pathogenic organisms. Reassurance can be obtained from colleagues involved in fertilization *in vitro*. The developing embryo is usually deposited into the uterus by a transcervical route. Infection has not been reported as a major cause of concern.

Reported incidence of infection after amniocentesis is about 3 per 1000, which is not too different from that observed for transcervical villus sampling (Brambati and Varotto 1985; Jackson and Wapner 1987). In the early stages of villus sampling prophylactic antibiotics or vaginal antiseptic pessaries were suggested (Brambati *et al.* 1985); but this has not been found to be necessary for all patients. Cumulative experience to date suggests risk of infection may be as low as 1 in 5000, and equates with that in the normal pregnant population (Jackson and Wapner 1987; Naeye

1983). Prophylactic antibiotics should, however, be prescribed for those with valvular heart disease, and should be considered in the more susceptible, such as diabetics and those on steroids.

Some slight bleeding often follows transcervical, and occasionally transabdominal, sampling, leaving a tract for possible ascending infection. It is sensible to take precautions, and we routinely obtain cervical swabs for chlamydia and viral and the other more usual pathogens a little before or just prior to villus sampling. Treatment of identified pathogens should prevent some of the late fetal loss following sampling, where infection may be implicated. Bacterial screening a long time before diagnosis is, however, not helpful, since the flora may change by the time sampling is performed (Brambati and Varotto 1985).

8.4 Ultrasound scanning

Ultrasound scanning and appreciation of the scan images obtained in the first trimester with various ultrasound machines are mandatory for villus sampling. Details of ultrasound machines, and their idiosyncrasies and usage are found in Chapters 6 and 7. Although many operators, at least initially, request assistance from ultrasonographers, the single-operator technique (Warren *et al.* 1987), where the obstetricians scan to direct their sampling, is considered ideal. Self-directed scanning improves co-ordination between the picture on the screen and placement of the sampling implement.

Scanning is performed before sampling to determine the viability, number, and normality of the pregnancy. Women referred for prenatal diagnosis are at increased risk of spontaneous miscarriage. Between 6 and 9 per cent of our patients referred for diagnosis miscarry or misabort before villus sampling is carried out. If multiple pregnancies are detected the couple must be recounselled. The risk of incurring miscarrage may be high, and thought must be given to the course of action when only one of the fetal constituents of the conceptus is abnormal. Selective fetocide does not appeal to all. The wrong fetus may be aborted, and any such attempt can induce miscarriage of the remaining pregnancy. If the decision is to proceed, insertion of the cord into the chorion frondosum must be identified, to avoid sampling from the wrong placenta. The site at the cord insertion is however vascular, and is usually best avoided, since haematoma formation may cause tamponade and fetal demise.

Abnormality of pregnancy must be identified, otherwise any potential miscarriage will be attributed to the sampling procedure. When there is discrepancy between crown–rump length and gestational sac size, or evidence of concealed bleeding, sampling should be deferred for a week to ten days. The diagnostic procedure is only carried out when normality is established at review. If missed abortion is confirmed, management will be along practised guidelines, with emphasis on support and much opportunity to discuss the whole issue with the couple concerned.

Following ultrasound verification of normality in the first trimester, the subsequent spontaneous-miscarriage rate up to 28 weeks is around 2.1 to 2.7 per cent for the normal population (Wilson *et al.* 1984; Liu *et al.* 1987). This miscarriage rate is increased to 4 or more per cent for the population of women referred for prenatal diagnosis (Wilson *et al.* 1984; Lind and McFadyen 1986). Risk of miscarriage decreases with pregnancy. At the end of the tenth week it is around 1.2 per cent, and remains at this level till 28 weeks (Liu *et al.* 1987). Villus sampling at this later stage of gestation, instead of the originally suggested eight weeks, will appear that much safer, since the spontaneous background loss is 2–3 per cent less. Sampling at this later stage may also be directly compared with that for amniocentesis, where the background loss-rate is similar.

The placenta is examined to determine the most suitable site for sampling. In practice the thickest area is selected. For the enveloping placenta of uniform thickness the direction of the cord insertion can be located to indicate the site of the chorion frondosum. Venous sinuses or lakes of various density are often seen (Fig. 8.1). If possible they should be avoided, to prevent contamination of the collected villi and the threat of haematoma formation and/or vaginal bleeding after diagnosis.

Once the implement is identified at the level of the internal os scanning is

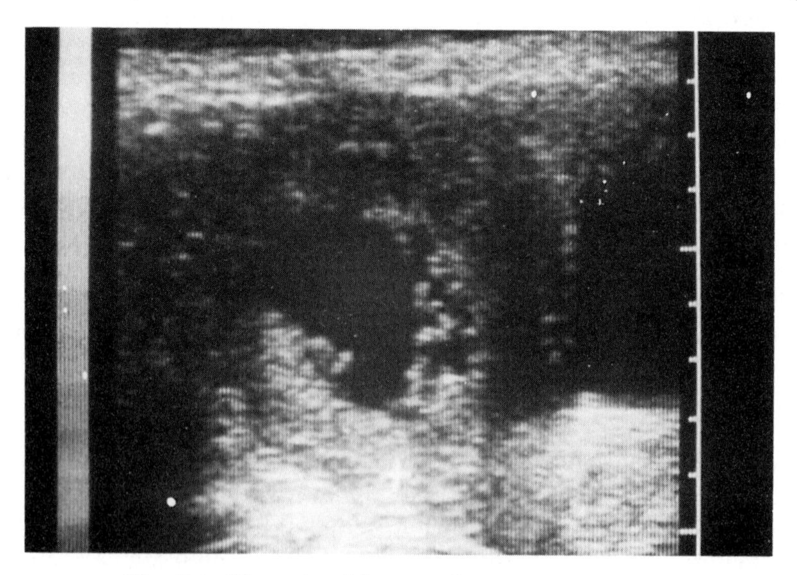

Fig. 8.1. Placenta with extensive venous sinuses.

continued to direct it to the chosen sampling site. Lateral movement should be limited, as this will cause unnecessary trauma or separation of the placenta. The path of the implement inside the uterus must be remembered, since withdrawal should follow as close as possible the path of insertion. Scanning should take into consideration the beam-width effect and the ultrasound artefacts described in Chapter 7. The ultrasound probe can be rocked to determine the exact position of the implement. Likewise rocking along the longitudinal axis with a linear or curvilinear probe will help delineate the tip of the sampling implements.

After sampling it is reassuring for the couple to see that the gestation sac is intact, the fetus is active, and the fetal heart is beating normally. Check the sampling path. If a haematoma is present, make sure it is not increasing, and review at weekly or appropriate intervals to follow resolution. Re-scanning is also sensible when complication develops.

8.5 Instruments

Requirements

Certain features are essential for any transcervical implement. It must be sufficiently long so that the fundal placenta of a 12 weeks' gestational uterus can be reached without the operator's hand encroaching into the vagina. Metal implements are usually more echogenic, and hence easier to locate and safer. The implement must be firm, to allow ready manipulation inside the uterus. This firmness also reduces the potential for bending and movement within the chorion—features inherent in implements made of plastic which are shown to be associated with a doubling of the fetal loss-rate (Anguo *et al.* 1985).

Manual dexterity and good hand–eye coordination are important assets conducive to a low intervention fetal loss-rate (Jackson and Wapner 1987). Likewise, if more than two insertions are needed to obtain a diagnostic sample this increases the hazards of the procedure (Jackson and Wapner 1987). Another requirement, therefore, is an implement which can secure sufficient diagnostic material at preferably two and at most three attempts.

Forceps

Those currently available are usually rigid, straight forceps seconded from paediatric usage, although a malleable version has recently been marketed (Fig. 8.2). The length is usually 25 cm, with a diameter of 2 mm. The tip is rounded for easy insertion, and when in position the jaws can be opened to capture a sample of villi (Dommergues and Dumez 1987). A requisite of

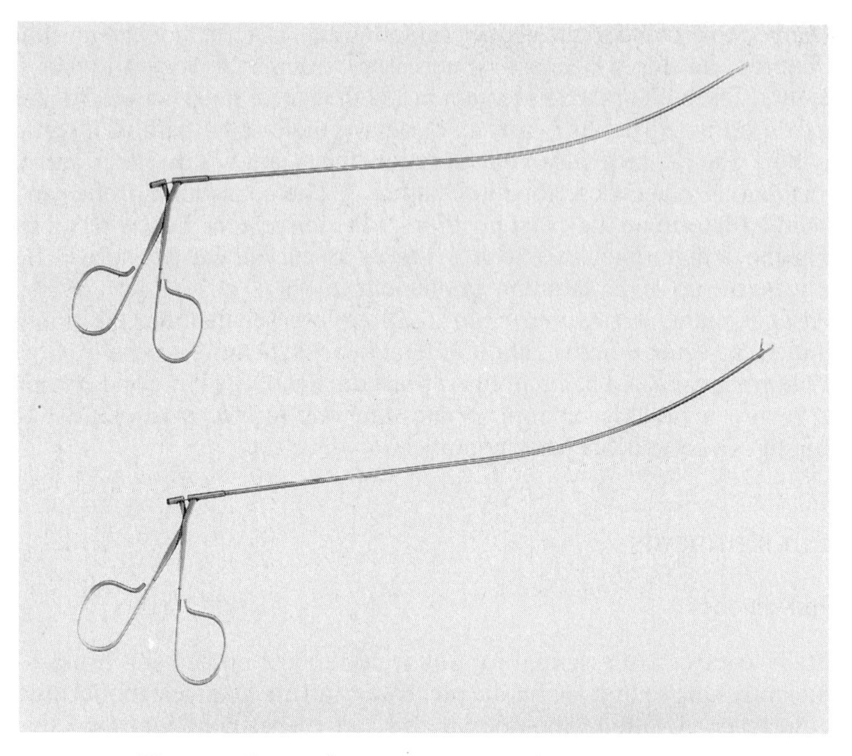

Fig. 8.2. Biopsy forceps (courtesy of D. J. Maxwell).

these authors' technique is application of a volsellum on the cervix to straighten the uterus to accommodate the straight forceps. All forceps are made of metal, and hence they are readily identified by ultrasound. Villus sampling is conducted as an out-patient procedure, without need for pre-medication. The pre-sampling preparations described previously are carried out. After placing the patient in the lithotomy position the external genitalia are cleansed with an antiseptic. Insertion of a speculum exposes the cervix for bacteriological screening and cleaning before application of the volsellum and passage of the forceps. The forceps are guided to the thickest part of the placenta by ultrasound. At the predetermined sampling site the jaws are opened, advanced a little, and then closed to grasp the villi before withdrawing. Successful sampling is indicated by observing a slight deformation of the chorionic plate, together with some resistance when the forceps are withdrawn. Adequacy of the sample is checked, and if necessary the procedure is repeated. The usual stipulate for the number of repeated insertions is imposed.

Sampling is performed between 9 and 11 weeks by the advocates of the use of this instrument. Lateral placentation presents most difficulty for forceps users (Dommergues and Dumez 1987). Failure to obtain sufficient diagnostic material occurs in between 0.5 and 1.0 per cent of cases. In some 2 per cent of patients a haematoma develops after sampling. To date the cumulative experience indicates the fetal loss-rate is of the order of 7.5 per cent, which is consistently higher than that found with other techniques (Jackson 1987, 1988).

Cannulae

These are usually made of plastic or metal. Metal cannulae are either rigid or malleable. One or more openings are usually placed at or towards the tip of the cannula. A source of negative pressure is required to capture and collect villus material. This may be by means of a foot-operated pump which delivers a continuous negative pressure, as used with the Down's cannula (Warren *et al.* 1987) or by means of a 5–10 ml syringe, which has been shown to be simple and safe (Liu *et al.* 1984). The author prefers use of the syringe, because once villus material enters the cannula it immediately becomes more difficult to withdraw the plunger, which thus gives an early indication of the likelihood of success.

Plastic cannulae

This is exemplified by the Portex (Portex Ltd, UK) cannula (Fig. 8.3). It is a single-use, flexible plastic cannula 21 cm in length and 1.5mm in external

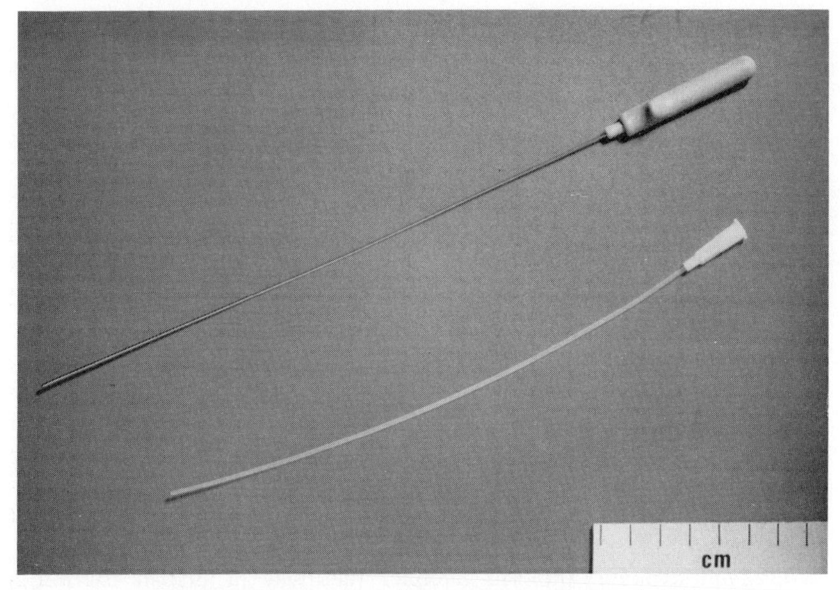

Fig. 8.3. Plastic cannula (courtesy Portex Ltd, UK).

diameter. A round-ended malleable metal obturator protrudes from the open end to exclude epithelium during insertion. The metal obturator increases echogenicity, and can be bent to assist delivery to the anterior placenta. For the posterior placenta the cannula is applied as for the anterior placenta, identified at the region of the internal cervical os, and then rotated 180° before advancing again. To steady the cervix a Babcock forceps is advised, since this instrument is less uncomfortable than tooth volsellums, and leaves little or no cervical bleeding. Once the cannula is in place the obturator is removed. This step of the procedure invariably causes movement and distortion within the chorion, and this leads to one of the main criticisms of plastic cannulae, namely the likelihood of doubling fetal loss (Anguo et al. 1985).

Metal cannulae

These implements are usually round-ended and approximately 2 mm in external diameter, with one or more openings close to the tip. Some are malleable, so that operators can fashion them into the shape they consider appropriate, while others, like the Liu Sampler (Femcare UK), have a fixed-curvature design to assist control and travel within the uterine cavity (Fig. 8.4). The latter cannula has two further features to enhance safety. An area

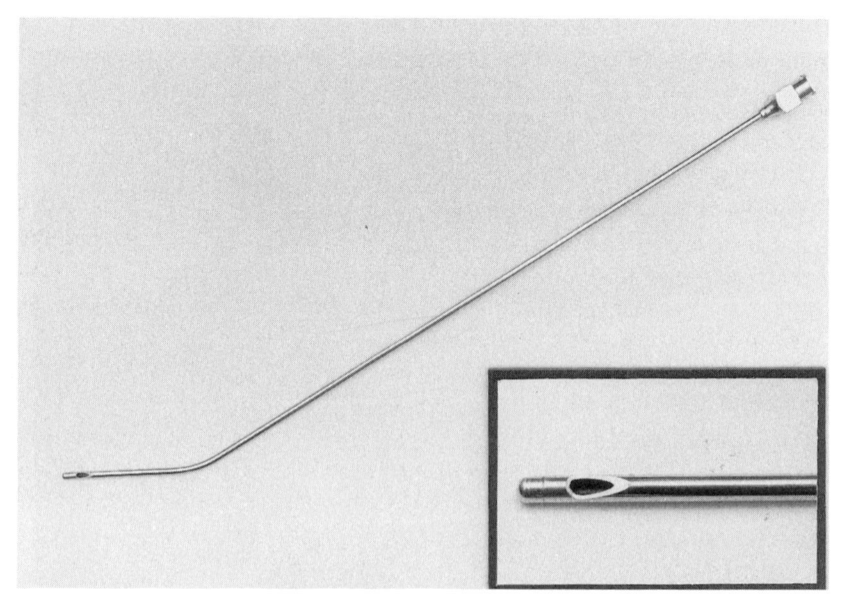

Fig. 8.4. Metal cannula: the Liu Sampler (courtesy of *British Journal of Obstetrics and Gynaecology*).

at the tip is ground flat to reflect ultrasound for easier cannula-tip localization. The single tear-drop opening with the point directed away from the tip is designed to cut the villi during aspiration, so lessening the to and fro movements suggested for villi collection. Furthermore, the opening is placed either on the left or the right, so that use of the appropriate cannula for lateral placentation avoids drawing in the amniotic sac membrane. Some metal cannulae are disposable, but others are sufficiently robust for repeated use—a feature not lost on those of us who must be cost-conscious.

In addition to being readily identifiable by ultrasound metal cannulae have the further advantage of being sufficiently firm to allow easy introduction through the cervix without assistance from uncomfortable cervical traction.

There are few studies which compare the utility of available cannulae. This may be because true comparison is difficult, since not many operators are equally familiar or practised with all these implements. Newcomers often adopt the preferences of their principals for their own preferred designs. Those interested in taking up chorion villus sampling are, however, advised to examine critically what is commercially available, and to select the cannula with the most design merits.

8.6 Sampling

Whichever cannula is used for transcervical aspiration of chorionic villi, the following steps are similar:

1. Check indication for and intent to undergo villus sampling is still valid. Obtain written consent.
2. Perform pre-sampling scan with a full bladder. Exclude the 6–9 per cent of referred patients who may have miscarried or misaborted when they present. Scan in both longitudinal and transverse direction. Determine the most appropriate site for sampling and the most suitable path towards it. Avoid venous sinuses if possible. Offer a new date if the stage of gestation is not suitable. Adequate follow-up and further counselling is important for patients with a miscarriage or misabortion.
3. The bladder can be emptied in a slim woman with an anteverted uterus, to make her feel more comfortable. A full bladder can also push the uterus out of the pelvis and out of easy reach. Conversely, a full bladder is helpful in displacing the loops of bowel from the uterus and enhancing visualization in the obese, or where the uterus is retroverted.
4. Place the patient in the lithotomy position, and drape and cleanse as for a surgical procedure. Insert speculum and take swabs from the cervix for

viral and bacterial culture if these were not previously obtained. Clean the cervix with an antiseptic solution.

5. Rescan the uterus, as the chosen site for sampling often changes when patients adopt the lithotomy position. Align the probe in the longitudinal axis to give a clear view from the internal cervical os to the sampling site.
6. Locate the tip of the cannula at the level of the internal os, before directing it to the sampling site (Fig. 8.5). Avoid lateral movements. Remember the direction of travel, so that withdrawal of the cannula follows the path of entry, so as to minimize trauma (Figs 8.6 and 8.7).
7. Aspirate with a 5 or 10 ml syringe once the cannula is in place. Some operators fill the canula with a sterile physiological solution, but this is not necessary. We do not advise use of heparinized solutions, since inadvertent spillage into the sampling site may encourage bleeding. Withdraw with a to-and-fro motion to assist the collection of villi. Warn

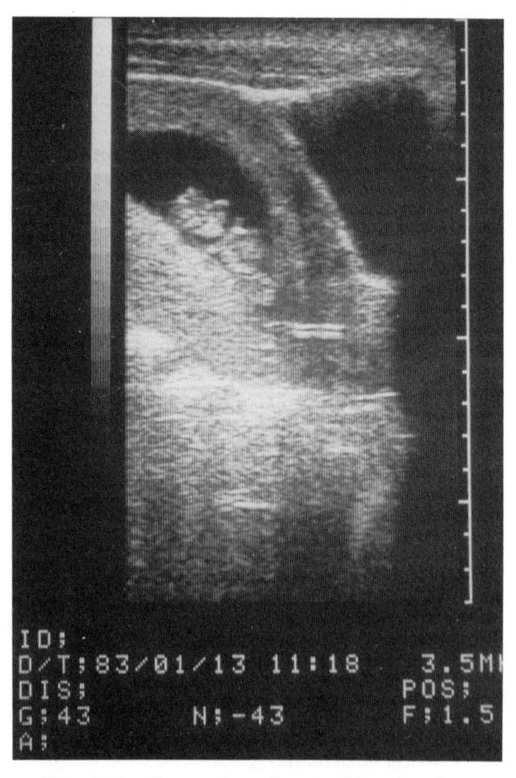

Fig. 8.5. Cannula at level of internal os.

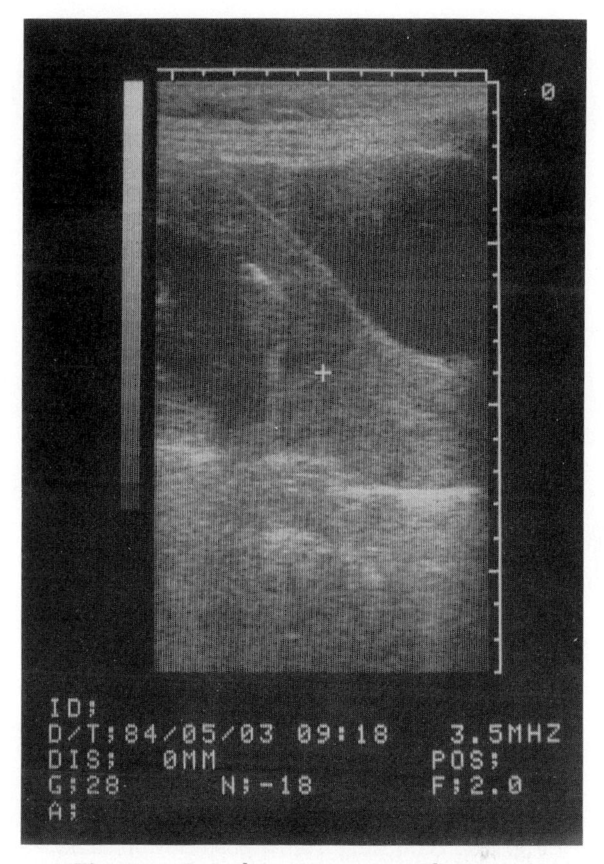

Fig. 8.6. Sampling an anterior placenta.

patients that a suction noise may be noticeable when the catheter is removed.

8. Place the sample in a Petri dish with culture medium, and examine it under a microscope. Remove blood clots and decidua if these are present. It is easy to recognize the type (villi with numerous buddings) and the amount (usually 10–25 mg wet weight) required for diagnosis (Figs 8.8 and 8.9). Blood present is usually recorded as 0–4 +, the latter notation indicating a specimen which is heavily bloodstained. Repeat the procedure if the specimen is inadequate or unsuitable.
9. Rescan to establish that the gestation sac is intact, the fetal heart is normal, and no haematoma is present (Fig. 8.10). Haematomas should be followed by repeat scanning to ensure resolution, usually by the sixteenth week (Brambati *et al.* 1987; Jackson and Wapner 1987).

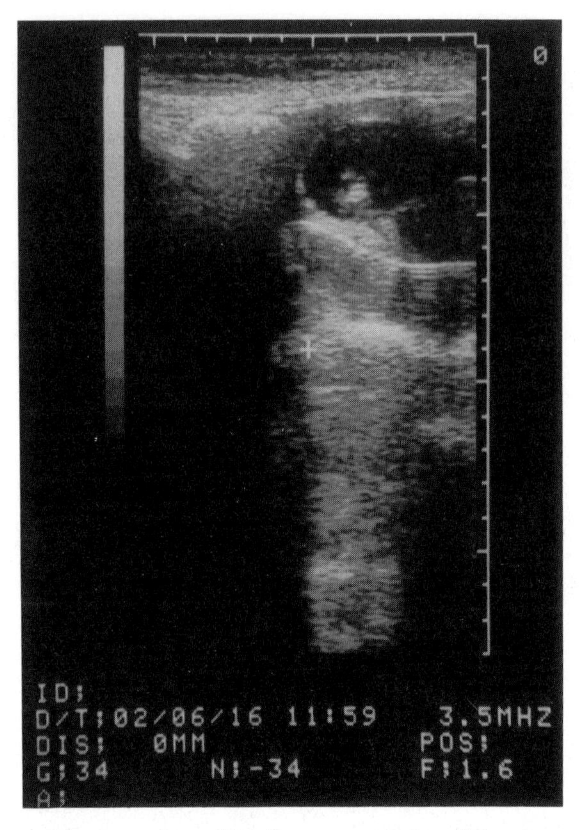

Fig. 8.7. Sampling from a posterior placenta.

8.7 Postoperative advice and management

Feto-maternal haemorrhage can occur, as shown by a rise in maternal serum alpha-fetoprotein (Brambati *et al.* 1986; Liu *et al.* 1983); but this bleeding is not detectable by the Betke–Kleihauer count (Warren *et al.* 1985; Liu *et al.* 1989). At risk Rhesus-negative women should receive 300 μg anti-D-immunoglobin to prevent sensitization. Whether sensitized women should undergo diagnosis remains controversial, though the same debate arises with amniocentesis. Repeated sampling, and a sample size of more than 50 mg wet weight, are more likely to provoke a rise in maternal serum alpha-fetoprotein (Warren *et al.* 1985; Brambati *et al.* 1987). This rise is transient, and begins to fall within 24 hours (Liu *et al.* 1989) to levels appropriate for gestation after 15 days (Perry *et al.* 1985; Brambati *et al.* 1986), so screening for neural tube defects should not be compromised.

Fig. 8.8. A sample of chorionic villi.

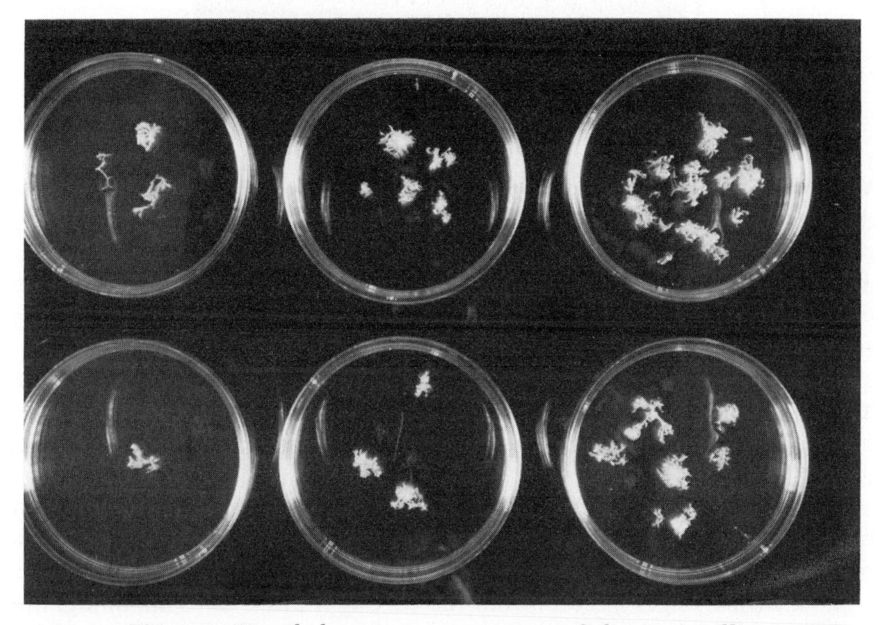

Fig. 8.9. Visual chart to assess amount of chorionic villi.

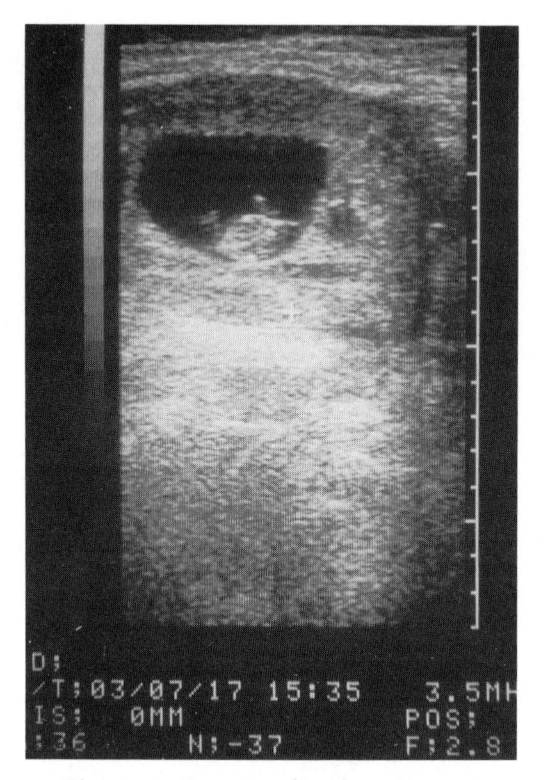

Fig. 8.10. Post-sampling haematoma.

Revealed bleeding, expressed mainly as spotting, is reported by 35–60 per cent of patients after villus sampling (Jackson and Wapner 1987; Warren *et al.* 1985). This may or may not be accompanied by a period-like discomfort. The patient should be warned of these likely occurrences, and reassured that the spotting should resolve with rest. Pregnancy outcome is not compromised (Brambati *et al.* 1987; Jackson and Wapner 1987) unless bleeding is heavy, when 25 per cent (1 in 4) to 50 per cent (4 out of 8) may subsequently miscarry (Jackson and Wapner 1987; Liu *et al.* 1989). Coitus should be avoided for the week after sampling and when spotting is evident. Miscarriage due to sampling is undoubtedly an important consideration; but the risk to the mother of a mid-trimester therapeutic abortion should not be forgotten. It is however becoming increasingly clear that in experienced hands with no more than two sampling attempts the intervention-related fetal loss-rate approximates to that for amniocentesis, and is possibly less than 1 per cent (Jackson and Wapner 1987).

Most miscarriages occur within the fortnight after villus sampling; but some late mid-trimester loss may be attributed to the diagnostic procedure.

Follow-up scans are advised, particularly where the sampling was considered difficult.

Occasionally intra-uterine infection, with clinical maternal symptoms, can occur (Blakemore *et al.* 1985). All patients should be advised to report back if they are unwell or there is pelvic pain and discharge. Prophylactic antibiotics should be considered in women with diabetes or heart-valve lesions.

8.8 Conclusion

Chorion villus sampling is an emergent technique, and as with any new procedure, good clinical practice necessitates careful documentation of any complications, to allow proper auditing for the benefit of subsequent participants. And again, as with all surgical techniques, once the skills are acquired they can only be maintained with constant practice. Chorion villus sampling is no different; there can be no place, in both the patients' or the technique's interests, for the occasional operator.

References

Anguo, H., Bingru, Z., and Hong, W. (1985). Long term follow-up results after aspiration of chorionic villi during early pregnancy. In *First trimester fetal diagnosis* (ed. M. Fraccaro, G. Simoni, and B. Brambati), pp. 1–6. Springer-Verlag, New York.

Blakemore, K. J., Mahoney, M. J., and Hobbin, J. C. (1985). Infection and chorionic villus sampling. *Lancet*, **ii**, 338.

Brambati, B. and Varotto, F. (1985). Infection and chorionic villus sampling. *Lancet*, **ii**, 609.

Brambati, B., Simoni, G., Danesino, C., Oldrini, A., Ferrazzi, E., Romitti, L., Terzoli, G., *et al.* (1985). First trimester fetal diagnosis of genetic disorders: clinical evaluation of 250 cases. *Journal of Medical Genetics*, **22**, 92–9.

Brambati, B., Guercilena, S., Bonnacchi, I., Oldrini, A., Lanzani, A., and Piceni, L. (1986). Feto-maternal transfusion after chorionic villus sampling: clinical implications. *Human Reproduction*, **i**, 37–40.

Brambati, B., Oldrini, A., Ferrazzi, E., and Lanzani, A. (1987). Chorionic villus sampling: an analysis of the obstetric experience of 1,000 cases. *Prenatal diagnosis*, **7**, 157–69.

Cuckle, H., Ward, N. J., and Thompson, S. G. (1987). Estimating a woman's risk of having a pregnancy associated with Down's syndrome using her age and maternal serum alpha-fetoprotein level. *British Journal of Obstetrics and Gynaecology*, **94**, 387–402.

Dommergues, M. and Dumez, Y. (1987). Chorion villus sampling using rigid forceps. In *Chorion villus sampling* (ed. D. T. Y. Liu, E M. Symonds, and M. S. Golbus), pp. 85–95. Chapman and Hall, London.

Garden, A. S., Reid, G., and Benzie, R. J. (1985). Chorion villus sampling. *Lancet*, **i**, 1270.

Jackson, L. (1987). *CVS Newsletters*. Number 2.

Jackson, L. (1988). *CVS Newsletters*. Number 3.

Jackson, L. G. and Wapner, R. J. (1987). Risks of chorion villus sampling. In *Clinical obstetrics and gynaecology. Fetal diagnosis of genetic defects* (ed. C. H. Rodeck), **1**(3), 513–31. Baillière Tindall, Great Britain.

Lind, T. and McFadyen, I. R. (1986). Human pregnancy failure. *Lancet*, **i**, 91–2.

Liu, D. T. Y., Mitchell, J., Johnson, J., and Wass, D. M. (1983). Trophoblast sampling by blind transcervical aspiration. *British Journal of Obstetrics and Gynaecology*, **90**, 1119–23.

Liu, D. T. Y., Slater, E., and Norman, S. (1984). Aspiration as a technique for biopsy of chorionic villi. *Journal of Obstetrics and Gynaecology*, **5**, 75–7.

Liu, D. T. Y., Jeavons, B., Preston, C., and Pearson, D. (1987). A prospective study of spontaneous miscarriage in ultrasonically normal pregnancies and relevance to chorion villus sampling. *Prenatal Diagnosis*, **7**, 223–7.

Liu, D. T. Y., Jeavons, B., Pearson, D., Preston, C., and Symonds, E. M. (1988). Patient experience of transcervical chorion villus sampling as an out-patient procedure. *Journal of Psychosomatic Obstetrics and Gynaecology*, **8**, 113–18.

Liu, D. T. Y., Preston, C., and Jeavons, B. (1989). Bleeding as a consequence of chorion villus sampling. *Asia–Oceania Journal of Obstetrics and Gynaecology*.

McFadyen, I. R., Taylor-Robinson, D., Furr, P. M., and Boustouller, Y. L. (1985). Infection and chorionic villus sampling. *Lancet*, **ii**, 610.

Naeye, R. L. (1983). Maternal age, obstetric complications, and the outcome of pregnancy. *Obstetrics and Gynaecology*, **61**, 210–16.

Perry, T. B., Vekemans, M. J. J., Lippman, A., Hamilton, E. F., and Fournier, P. J. R. (1985). Chorion villi sampling: clinical experience, immediate complications and patient attitudes. *American Journal of Obstetrics and Gynaecology*, **151**, 161–6.

Warren, R. C., Butler, J., Morsman, J. M., McKenzie, C., and Rodeck, C. H. (1985). Does chorionic villus sampling cause feto-maternal haemorrhage? *Lancet*, **i**, 691.

Warren, R. C., McKenzie, C. F., Kearney, L., and Rodeck, C. H. (1987). Transcervical chorion villus aspiration by a single operator technique. In *Chorion villus sampling* (ed. D. T. Y. Liu, E. M. Symonds, and M. S. Golbus), pp. 65–72. Chapman and Hall, London.

Wass, D. and Bennett, M. J. (1985). Infection and chorionic villus sampling. *Lancet*, **ii**, 338–9.

Wilson, R. D., Kendrick, E., Wittman, B. K., and McGillivray, B. C. (1984). Risk of spontaneous abortions in ultrasonically normal pregnancies. *Lancet*, **ii**, 920.

9 Transabdominal chorion villus sampling

D. J. Maxwell

9.1 Introduction

The transabdominal technique for chorion villus sampling (CVS) (Smidt-Jensen and Hahnemann 1984; Maxwell *et al.* 1986) has in recent years become firmly established as an alternative to the transcervical approach. Indeed, in some centres with the experience of both techniques, it has come to be the preferred method.

Reasons why the transabdominal route could be considered more appropriate for CVS include an avoidance of the cervical canal, with its uncertain microbiological composition, and the more direct route to the target tissue, requiring less intra-uterine manipulation. An additional practical reason for suggesting the transabdominal as the primary route is that of its fundamental similarity to the other prenatal diagnostic techniques of amniocentesis and cordocentesis. When operators perform any one of these techniques they are effectively refining their ability to perform the others.

9.2 Instrumentation

Scanning

An ultrasound machine with good resolution is essential to the performance of CVS whether it is performed transabdominally or transcervically. A sector or curvilinear transducer is necessary, the spread of the beam allowing reliable visualization in the first trimester. The curvilinear transducer is also easier to apply to sampling attempts later in pregnancy. Linear probes are generally unsuitable. I believe it is most appropriate for operators to do their own scanning while performing these techniques, since it is essential to have the sampling needle under continuous vision at all stages. This can be best achieved by co-ordination of the same pair of hands, one holding the sampling needle and the other making the minute adjustments to the scanner position that are necessary for continuous visualization. The operator must be adept in the use of ultrasound. This of course, is eminently desirable, since the people performing the transabdominal CVS are the most likely to be involved in other prenatal diagnostic techniques in their hospitals. A very experienced and successful combination of ultrasonographer and obstetrician can also be achieved, with each knowing what the other is attempting to do. The obstetrician in these circumstances invariably has more than a fundamental knowledge of the limitations of ultrasound. There is no place for the occasional operator.

Approaches

There are two fundamental approaches to transabdominal chorion villus sampling:

(1) needle-guided; and

(2) freehand.

I have now adopted the freehand method for all cases, having initially used a needle-guided system.

The essential requirement for the guide is that it should allow easy passage of the sampling needle through it while keeping the needle firmly within the ultrasound beam. Most guides of commercial manufacture are inadequate in the later respect. They allow excessive movement of the needle, with the result that the tip of the needle can swing through a wide arc, and can be deflected from within the ultrasound beam during passage through the maternal soft tissue. The guides that I have used in the past have all been made in hospital workshops. If the ultrasound machine has within its software a needle biopsy guideline, then the needle entry through the guide can be adjusted during the manufacture to coincide with this. Alternatively, with the guide attached to the transducer and the needle in place, the assembly is dipped into a waterbath. The strong echoes from the needle are easily seen, and can be plotted on a plastic template which can then be placed over the viewing screen whenever the guide is to be used. Both methods work well. A needle-guide offers some advantages, and may be easier to use initially. The end-result is that whenever the guide is properly attached to the transducer and the guidelines are overlaid on the screen, the intended needle path is outlined, allowing scanning for an ideal entry-site. The ease or otherwise of sampling is readily anticipated, and sampling should not commence until success can confidently be predicted.

The close proximity of transducer, guide, and needle requires that the transducer be sterile, as are the guide and needle itself. This is achieved by encasing the transducer in a sterile polythene bag or glove with ultrasound gel on the inside. As has been said, the needle-guide system has worked exceptionally well, with a 100 per cent diagnostic sampling success-rate in my experience.

Disadvantages of a needle-guided system are readily apparent, particularly in retrospect. The manufacture of the guide requires a marked commitment from the workshop and the intending user. The assembly of the components at the beginning of each procedure takes time, and brings a level of complication to what is essentially a simple procedure. Since the transducer and needle are locked together, independent movement of the transducer is not possible. This is a particular disadvantage when the best scanning site and the ideal needle entry-position are remote from each other, as may occur with very shallow angles of entry. The freehand approach avoids all these disadvantages. It is rather more exacting of the operator's skills, which may be an advantage to the patient.

Sampling devices and technique

The choices of sampling device for transabdominal CVS are:

(1) co-axial needle combination and syringe suction;

(2) single needle and syringe suction; and

(3) co-axial combination with biopsy forceps.

The co-axial needle combination is that of a 17-gauge needle with stylet which has been specially bored out to accept a 19-gauge aspirating cannula. Since the inner 19-gauge aspirating cannula is 2 cm longer than the outer guide, it projects further into the placenta (see Fig. 9.1a). Syringe suction and gentle up-and-down movement allow villi to be aspirated. An alternative combination is that of 18 and 20 gauges. Both sets of needles are easily available from a number of suppliers, and are relatively inexpensive.

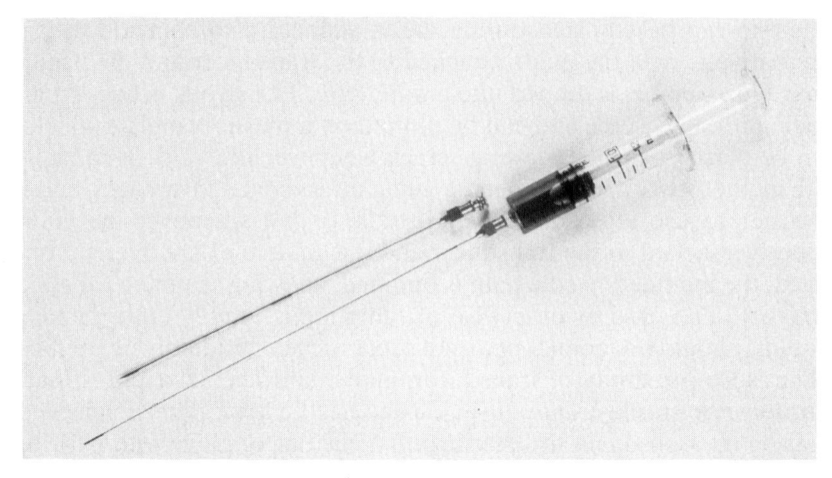

Fig. 9.1a. Transabdominal chorionic villus sampling needles. The aspirating cannula with syringe attached is 2 cm longer than the guide.

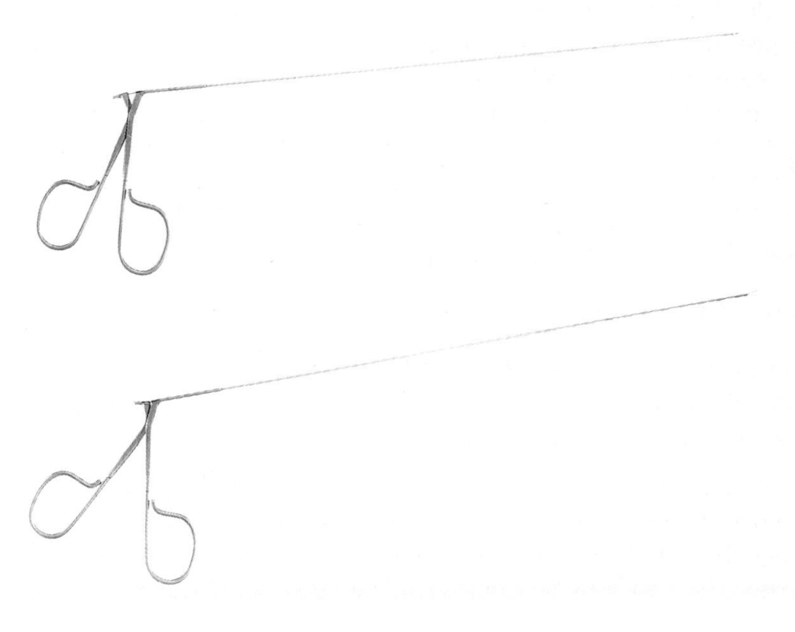

Fig. 9.1b. Biopsy forceps to be used co-axially with a guide for transabdominal chorionic villus sampling.

An initial purchase of six sets to be sterilized and re-used in rotation is advised. In case (3) tiny biopsy forceps replace the inner cannula (Fig. 9.1b). Co-axial combinations have the advantage that multiple sampling attempts can be performed for a single insertion of the guide-needle. If more villi are required, then it is easy to reinsert the aspirating cannula/forceps through the 17-gauge needle, which is only withdrawn once sampling adequate for diagnostic purposes is completed. This has a particular advantage when more than one study is intended on the villi, for example karyotyping plus DNA studies. With syringe suction generally only one or two aspirations are necessary to allow 20 to 40 mg of tissue to be collected. Each additional aspiration involves no further discomfort to the patient. A single 20-gauge needle with syringe suction is used by Brambati and colleagues in Milan (Brambati *et al.* 1987). They have considerable experience of both transcervical and transabdominal CVS. The smaller needle is used to achieve a reduced procedure-related pregnancy-loss. This is as yet unproven. The single-needle yield gives much the same quantity of villi as the co-axial system, but requires the needle to be reinserted through the patient's abdomen if a further sample is needed. If too little material is obtained, more pressure may be placed on the laboratory to work with a suboptimal sample to avoid a second skin insertion—a potential disadvantage which they have been able to avoid. Having evaluated both systems, I have found the co-axial needle combination to be an extremely effective system, combining minimal patient discomfort with maximum laboratory satisfaction.

A co-axial combination using biopsy forceps is possible, and our initial experience is favourable. Replacement of syringe suction by biopsy forceps results in cleaner specimens. However, more insertions (between 4 and 6) are necessary, because of the inherently small size of the samples. The principal disadvantage of this system is that the procedure time is longer.

Care of instruments

The instruments are sterilized by autoclaving or Cidex immersion. The sampling needles do need to be adequately protected during the autoclaving process and in transit. Glass vials serve these needs very well. A matched set of guide plus aspirating cannula is contained in each vial, with a plug of cotton wool at the end to protect the tips. The vials provide an ideal method of storage and transport, and can be easily labelled to prevent them going astray. Suitable alternatives are cardboard trays or plastic sheaths for the needles, and then disposable paper wrappings. If four or six needle-sets are purchased they can then be used in rotation. The initial cost of this number of sets is not excessive. If a needle-guide is employed, it is most important to autoclave it between sessions, and to immerse it in a Cidex solution for 20 minutes between patients. Needles can also be placed in Cidex as an alternative to autoclaving if fewer sets are used. Before use they should be rinsed with sterile water or normal saline to remove excess Cidex, which is irritant to soft tissues.

9.3 Freehand transabdominal sampling

The freehand approach to transabdominal CVS will be described in some detail, as this is now the method I always use. Transabdominal CVS can be divided into a number of stages. An initial scan is performed to establish fetal viability and gestation, and to establish the accessibility of the villi. Bladder-filling is adjusted if necessary (see Figs. 9.2 and 9.3). This initial assessment is the most important phase of the technique. Success is dependent upon an unhindered route displayed on the ultrasound screen. Absolute consideration to this point will be repaid by a reliable sampling rate. Once a straight line of entry through the abdominal wall and the uterus and into villi parallel to the uterine cavity can be seen, then sampling may begin.

The preparation time for freehand transabdominal CVS is minimal. A small disposable pack which contains a plastic gallipot and cotton wool

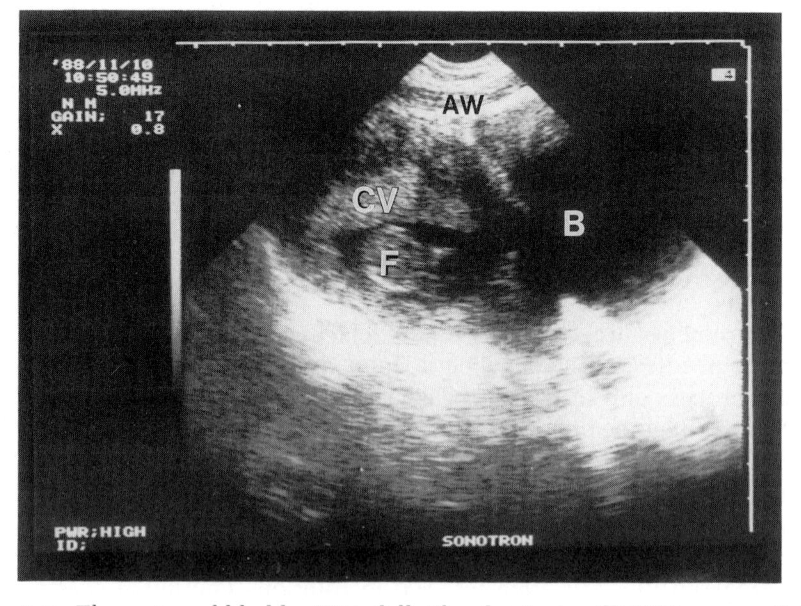

Fig. 9.2. The maternal bladder (B) is full. The chorionic villi (CV) are parallel to the underside of the maternal abdominal wall (AW). The fetus is shown at (F).

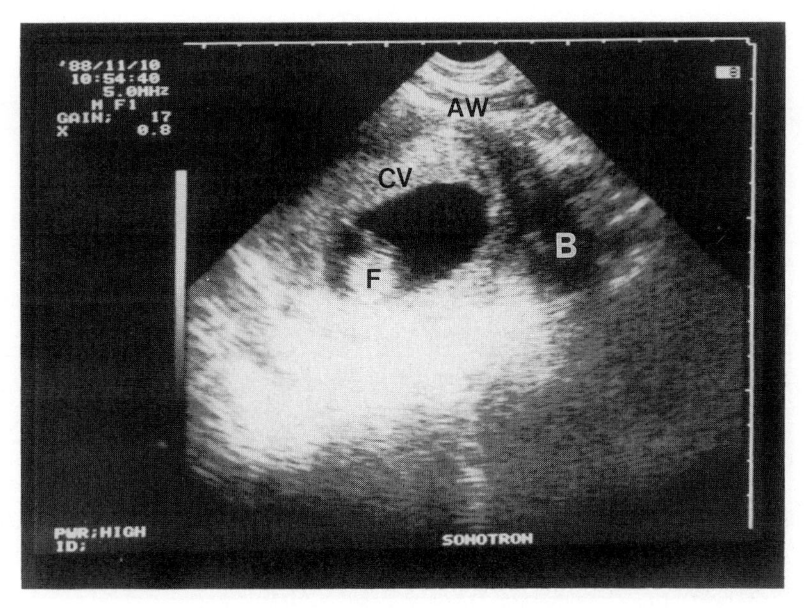

Fig. 9.3. Bladder (B) has been emptied. The chorionic villi (CV) are now accessible via the abdominal wall (AW). The fetus is shown at (F).

balls is opened. Two 10 ml syringes are then dropped on to the pack, together with a needle for administering local anaesthetic and the combination needles. Gloves are donned, but not gown or mask. The needle combinations should always be checked before proceeding, as they can sometimes be mismatched during sterilization. Local anaesthetic is drawn up, and aqueous Betadine is placed in the gallipot. A 10 ml syringe is attached to the 19-gauge sampling cannula, and 3 to 5 ml of wash or culture medium are aspirated into this system. The ultrasound transducer is then held in the left hand, the right remaining sterile. The target villi are rechecked, and the entry site is re-chosen. The skin for an area of about 10 cm around the proposed puncture site is cleaned, and then local anaesthetic is instilled into the skin and abdominal wall. Once this has been done, the local anaesthetic needle can be advanced until it appears on the ultrasound screen. This step allows the intended angle of entry of the first needle to be checked in relation to the villi (Fig. 9.4). Care is taken not to insert local anaesthetic into the myometrium, as this would cause uterine spasm and might allow entry of local anesthetic into the maternal circulation. The 17-gauge guide-needle is then inserted through the abdominal wall in the path of the ultrasound beam, and can be advanced under view towards the uterus. Once the needle can be seen in line with the villi, it is then inserted through the uterus with a swift controlled motion designed to bring the

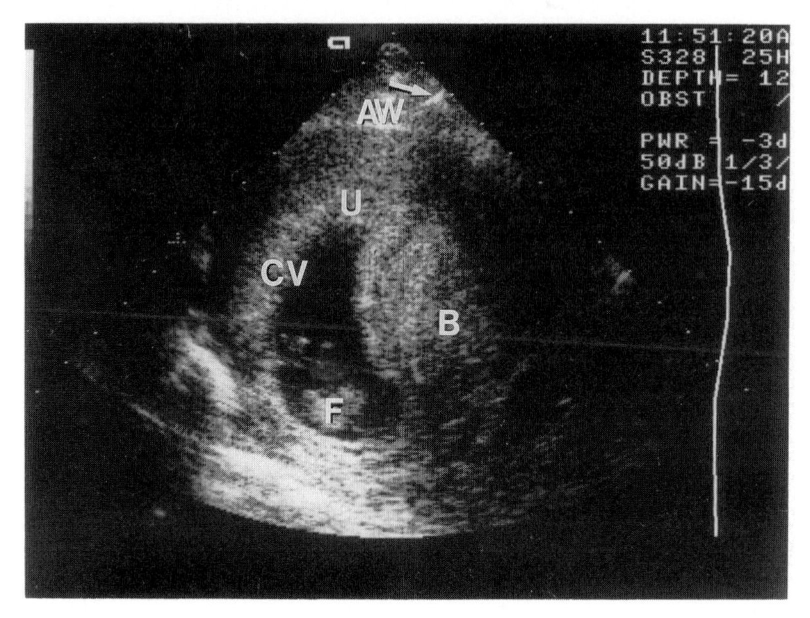

Fig. 9.4. The maternal bladder (B) is empty. The fundus of the uterus (U) is in contact with the underside of the abdominal wall (AW). The local anaesthetic needle (arrowed) is used to check the entry line to the chorionic villi shown at (CV).

point well within the substance of the villi. The needle-tip can be seen as a bright dot clearly visible among the less echogenic villi. The stylet is withdrawn, and the 19-gauge cannula, filled with wash medium and attached syringe, is then introduced. The wash medium makes the aspirating cannula even more echogenic, and the second tip can be seen lying 2 cm further into the villi than the first (see Fig. 9.5). Suction, together with a gentle up-and-down movement, is then performed to obtain villi. I ask an assistant to hold the outer needle gently while suction is occurring, in order to prevent the 17-gauge needle from being displaced. Alternatively, the assistant can perform the aspiration while the guide-needle is held by the operator. It is important to have the needle-tip under view while sampling is taking place, in order to avoid damge to the membranes or inadvertent aspiration or decidua. Aspiration continues for 10 to 15 seconds; the inner cannula is then withdrawn, and the villi are expressed into a receiver for examination under microscopy. It is a wise precaution when doing this to use a sterile universal container rather than a Petri dish. If, as sometimes happens, the villi impact in the aspirating cannula, the extra pressure necessary to expel them may cause splashing out from the shallow sides of the Petri dish, with consequent loss of specimen. An assessment of villus quantity and quality

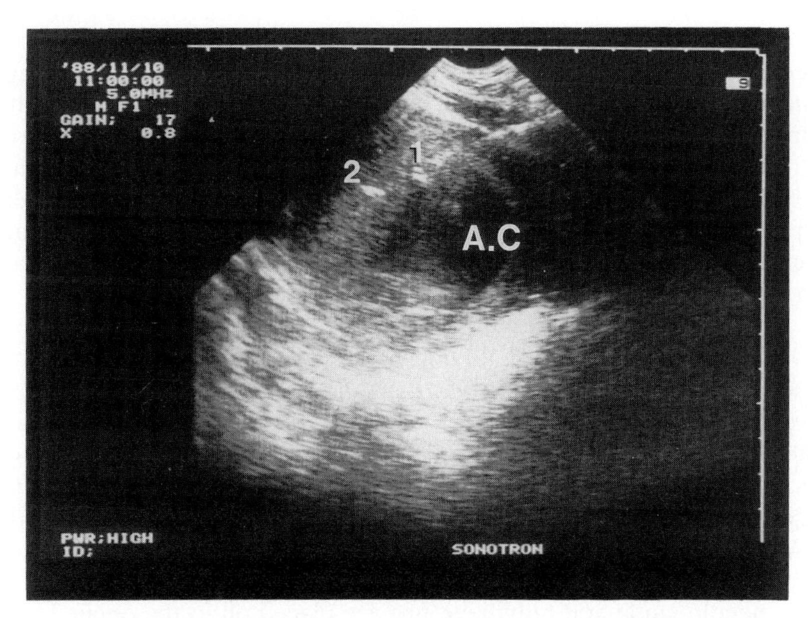

Fig. 9.5. Sampling is taking place. The path taken by the guide-needle can be seen clearly, with its tip at 1. The end of the aspirating cannula is at 2, well within the substance of the chorionic villi. The amniotic cavity is shown at (AC) (gestation 10 weeks).

can be given immediately. If more villi are required, the aspiration cannula is refilled with wash medium and reinserted through the 17-gauge guide-needle, which has remained *in situ.* If a second aspiration is performed, enough lateral movement of the guide and aspirating cannula can be obtained to sample from a separate area a small distance away from the original site. The needle is in the patient's abdomen for an average of two minutes per procedure.

The procedure seems to be well-tolerated by the vast majority of patients. All are properly counselled beforehand. In addition, I repeat an explanation of what they can anticipate before starting, and talk them through each step of the procedure. A small amount of inadvertent movement by the patient as the uterus is entered is common. This is potentially the most uncomfortable part of the procedure, and it is wise to warn the patient immediately prior to penetration. Once the 17-gauge needle is *in situ* the remainder of the sampling is relatively painless.

Freehand transabdominal CVS will allow a high patient throughput if desired. They are asked to attend with a moderately full bladder. They do not have to disrobe, but simply lie on an adjustable height couch. If a satisfactory line of entry can be seen on scan, the procedure can be commenced within a few minutes, and completed soon after. Alternatively, the patient is asked to empty or fill her bladder prior to rescan. While bladder filling is taking place, another patient can be seen. The lack of need for the patient to undress or to adopt the lithotomy position and the speed with which preparation can take place lead to an efficient use of facilities. Six patients can be easily accommodated in a three-hour session, and we have often seen nine. We prefer counselling to have taken place on a separate day from the procedure. However, this is not invariable, particularly if patients have to travel long distances or have severe constraints on their time.

9.4 Needle-guided transabdominal sampling

The preparation time for a needle-guided approach is a little longer. Uterine position is adjusted, as for the freehand approach, by adjusting bladder emptying. The ideal line of entry will be displayed by the needle-guide line on the ultrasound picture. A small pack is opened and gloves are donned. In addition to the items mentioned for the freehand approach, the sterile needle-guide and polythene bag are placed on to the opened pack. Ultrasound gel is placed in the end of the bag, and the transducer then covered by it, and the needle-guide is attached to the transducer (Fig. 9.6). Skin preparation is performed, and sterile ultrasound gel is placed on the abdomen. Further scanning is performed to check entry position. Once the

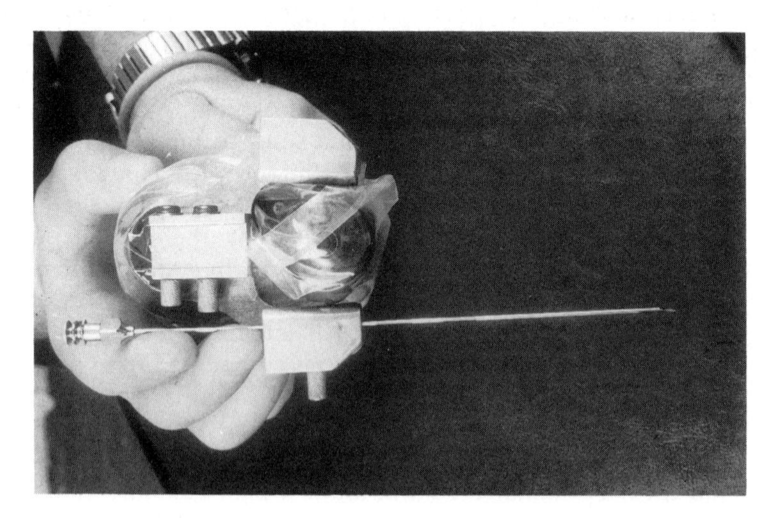

Fig. 9.6. Needle-guide attached to transducer. The transducer is covered with a polythene bag, which can be pre-sterilized. The needle can be passed down either side of the transducer, as desired.

transducer has been positioned, it is then held steady while the skin is infiltrated with local anaesthetic through the needle-guide. The 17-gauge needle is then inserted. In a properly constructed system, the needle will exactly follow the guideline on the ultrasound screen, and will be under vision at all times (Fig. 9.7). Entry is by swift controlled motion, to prevent the needle-tip from being displaced by maternal soft tissue. The remainder of the procedure is as for the freehand approach. Either a single needle or a co-axial combination can be used. Once again, it should be stressed that the sampling attempt should not take place until the ultrasound picture and proposed line of entry are clear.

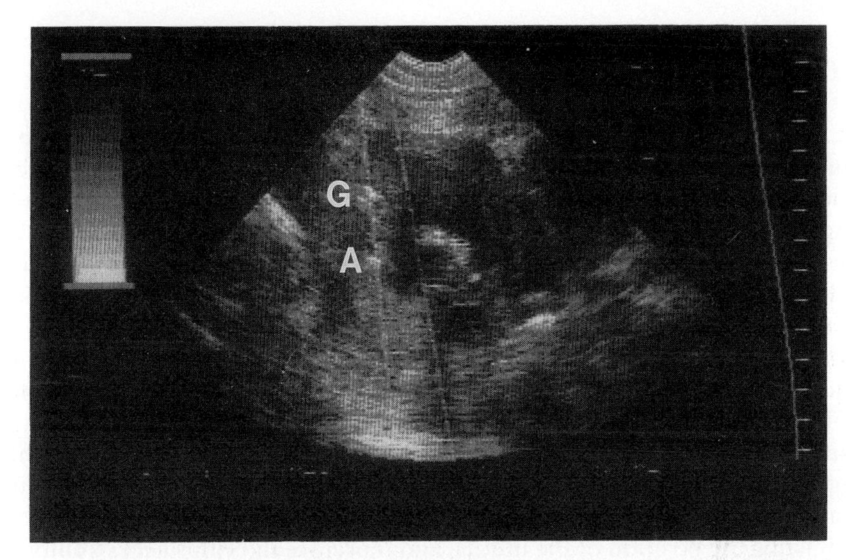

Fig. 9.7. The guide-needle has been passed through the needle-guide and along the sampling line, with its tip shown at (G). The aspirating cannula (A) projects further into the villi.

Comparing the two techniques, the freehand method is superior, if a little more exacting. It is simpler and allows a more flexible approach to villus positions; the preparation time (during which subtle changes in uterine positions may occur) is less; and the whole technique is identical to that adopted for any other prenatal diagnostic or therapeutic ultrasound-guided procedure.

9.5 Results

I have now performed 237 transabdominal chorion villus samplings for a wide spectrum of clinical indications. Tissue adequate for sampling has been obtained in all patients. Of 195 patients in whom the pregnancy was intended to continue, there has been only one loss—an intra-uterine fetal death diagnosed five weeks after an uncomplicated sampling. Comparable figures from the World Registry of Chorionic Villus Sampling suggest a 2.49 per cent fetal loss-rate from 5012 patients sampled (Jackson 1989). Experience from reporting centres who have large diagnostic series have shown a reasonable uniformity in pregnancy loss-rate.

It is anticipated that when our numbers are much greater, we will encounter a similar level of procedure risk. For the moment however, our own figures are encouraging.

9.6 General considerations

In performing transabdominal CVS an important point to realize is that conventional descriptions of placental position—that is, as anterior, posterior, anterofundal, etc.—are misleading and, I believe, irrelevant when applied to this technique. The fundus of the first trimester anteverted uterus comes into direct apposition with the inner surface of the anterior abdominal wall as the maternal bladder empties. In this situation, placental positions described as anterior or posterior come to lie with the long axis of the villi at right angles to the underside of the abdominal wall, and therefore equally accessible to a needle passed through the abdomen. Proper nomenclature for placental position relates its position to the position of the internal cervical os. In this first trimester this is often not readily apparent on scan. Patients are asked to attend with a moderately full bladder, which can be emptied in stages until the line of entry is seen. Adjusting the position of the uterus, and therefore the villus accessibility, is fundamental to sampling success. The single most important item in successful sampling is the ultrasound picture that is obtained prior to the commencement of the procedure, not placental position. On this picture, the absolute requirement is of an unhindered, preferably straight passage of the sampling device to the target villi (Figs. 9.3 and 9.5). If this requirement is fulfilled, for an experienced operator success is then guaranteed. If it is not, sampling should not commence until it is. Two patients have been asked to return a few days later when an ideal view was not obtained at first attendance. The sampling when performed was straightforward.

A common misbelief is that the combination of a retroverted uterus and a posterior placenta is a difficult situation to sample transabdominally. In practice this is not so. Invariably in the first trimester, a portion of the villi will not be seen laterally that can be reached by the sampling needle. Additional pressure of the transducer on the maternal abdominal wall may be necessary to ensure a good view. It must be stressed again that conventional descriptions of placental position are not helpful, since the only criterion to be fulfilled in this regard is an area of villi long enough to accommodate the aspirating cannula and directly accessible through the maternal abdominal wall.

Transabdominal CVS can be performed at any gestation from nine weeks onwards. It has an advantage over transcervical CVS in that sampling can be performed beyond the first trimester (Nicolaides *et al.* 1986). There are circumstances under which villi for laboratory analysis are preferable to amniotic fluid or fetal blood sample in achieving a diagnosis in later pregnancy—for example, karyotyping within 24 to 48 hours. Technically, sampling in the second trimester is often easier than in the first. However, it should be remembered that beyond about twelve weeks of pregnancy a true posterior placenta may not be reached directly without traversing the amniotic cavity. It is not clear whether transamniotic transabdominal CVS carries a higher risk to the pregnancy. Inadvertent entry of the amniotic cavity with the sampling needle is a potential complication of the technique. In four cases in our series where the amniotic cavity has been entered, deliberately or by accident, no pregnancy-loss has resulted.

Difficulties may be encountered in a number of clinical situations. Maternal obesity can be sufficient to interfere with the quality of the ultrasound picture, making the procedure more difficult. Fibroids within the uterus may appear to block the intended sampling line, and uterine contractions may alter villus accessibility. As has been emphasized earlier, restraint is required. Waiting for a short time, or even deferring the procedure by a few days, will invariably result in an ultrasound view which will allow sampling to proceed. A transvesical approach has been used on one patient. She was very obese, and a transcervical attempt had failed previously. The obesity precluded a proper ultrasound view with anything other than a very full bladder. An uncomplicated term pregnancy followed successful sampling.

Trauma to other intra-abdominal organs is possible. Bowel between the abdominal wall and the fundus of the uterus will interfere with ultrasound viewing, and should therefore be easily recognized. Adjusting bladder filling/emptying or additional pressure with the transducer will displace it from view. As long as an unobstructed view of the line of needle-entry is seen, bowel trauma should not occur. Involuntary maternal movement might, in theory, cause this complication. An appropriately sterile technique should not allow the possibility of maternal infection to be more of a problem than with amniocentesis.

Transabdominal CVS undoubtedly involves more patient discomfort than the transcervical approach. The degree of discomfort is variable, but seems well tolerated by most patients. Discomfort is maximal at the point of myometrial entry, and the actual sampling is almost painless. Additional quantities of villi can be obtained without the necessity for repeated transcervical negotiations, which can be uncomfortable, and may increase the risk of the procedure. Understandably, some patients fear needles to a marked degree. Involuntary movement during the course of the procedure, especially at myometrial entry, is a real problem, and has led to incorrect placement of the sampling needle on four occasions. A second skin-puncture has been successful in each case. If such maternal apprehension can be predicted beforehand, and transcervical CVS is an inappropriate alternative (for example, in cases of a failed previous transcervical approach or of gestation beyond twelve weeks), a small dose of intravenous analgesia given prior to sampling seems to overcome this problem. We have administered this to two patients. Adequate explanation, a relaxed patient, and good communication during the procedure will prevent or minimize the problem. Chorion villus sampling at Guy's Hospital is offered within the overal structure of the Medical Research Council Trial. Patients allocated to CVS are randomized to either the transabdominal or the transcervical route. We are currently evaluating patient-responses to the method of CVS that they undergo. Patient-acceptance of both methods is very similar in our study to date.

Transabdominal CVS is perceived by some obstetricians as being too sophisticated a technique to be available outside specialized centres. I do not share this view. Initial training is best done on consenting women undergoing therapeutic termination of pregnancy. However, constraints on time and the availability of scanning facilities in the operating theatre may place severe practical limitations on this method of training. It should be remembered that amniocentesis is best performed by a simultaneous scanning technique in which the needle is under continuous vision at all stages of entry through maternal soft tissue and into the amniotic cavity. This same technique is fundamental to freehand transabdominal CVS. It follows that anyone regularly performing amniocentesis in this fashion has it within their ability to perform transabdominal CVS. Attendance at an established centre initially, boosted by amniocentesis training, should allow a rapid learning curve on theatre cases. If one obstetrician from a group were to undertake all prenatal diagnostic procedures within that centre, then his or her ultrasound-guided needling expertise could be maintained at an acceptable level. Under these circumstances, transabdominal CVS becomes the logical procedure of first choice for first-trimester villus sampling.

9.7 Conclusion

Transabdominal CVS has proved to be a valuable method of prenatal investigation, enjoying high sampling success, simplicity of procedure, flexibility of approach, and good patient-tolerance. I have found the double-needle system with an ultrasound-guided freehand approach to be the most efficient method of obtaining villi, and therefore highly recommend it. Available evidence supports the relative safety of the procedure. Similarity with other prenatal needling techniques suggest the logic of the approach, and assists in the training of future operators. The adoption of this technique need not necessarily be restricted to major centres.

References

Brambati, B., Olrini, A., and Lanzani, A. (1987). Transabdominal chorionic villus sampling: a freehand ultrasound-guided technique. *American Journal of Obstetrics and Gynaecology*, **157** (1), 134–7.

Jackson, L. (1989). *CVS Newsletter*, 31 January.

Maxwell, D. J., Lilford, R., Czepulkowski, B., Heaton, D., and Coleman, D. (1986). Transabdominal chorionic villus sampling. *Lancet*, **i**, 123–6.

Nicolaides, K. H., Soothill, P. W., Rodeck, C. H., and Warren, R. C. (1986). Why confine chorionic villus (placental) biopsy to the first trimester? *Lancet*, **i**, 543–4.

Smidt-Jensen, S. and Hahnermann, N. (1984). Transabdominal fine needle biopsy from chorionic villi in the first trimester. *Prenatal Diagnosis*, **4**, 163–9.

10 Establishing a service for chorion villus sampling

David T. Y. Liu

10.1 Introduction

Chorion villus sampling is essentially a simple procedure, well within the capability of most obstetricians. In distinction to the situation in amniocentesis, however, the first-trimester uterus presents a much smaller target, which is less forgiving of mistakes, and this necessitates a longer training period. Adequate training is mandatory, and cannot be overemphasized. The simplicity of the procedure can be deceptive, and over-enthusiasm results in some of the reported high fetal wastage of 30–50 per cent. This wastage is detrimental to everyone's confidence in the test, and undermines the development of what is acknowledged as a step in the right direction of obstetric progress. The aim of this chapter is to take the reader through the seldom-discussed steps of training and the organization of support facilities needed to provide a service.

10.2 Subjects for training

Women requesting therapeutic abortion are usually approached to assist with a training programme. Incorporation of this category of women into any study is a sensitive issue. Strict adherence to a considered structured programme is advised to avoid criticism. We found the following steps helpful:

1. Approach only those women who are aged sixteen or over, and are conversant in the language used, so as to be able to give valid informed consent, and who are not unduly distressed.
2. Counselling for participation should only be given after the woman has been accepted for the therapeutic procedure. This will avoid any criticism that these women have been coerced into participating. As an added precaution we use a research sister for this purpose, to impose a physical division between the obstetrician's responsibilities and the trainer's interest in the patient's contribution.
3. Any infringement on the fetus or fetal tissue invariably evokes anxieties in both patients and medical staff. There are clear guidelines for handling fetal material (Peel Report). The rationale for use of the trophoblast and the mode of deployment of the tissue obtained must be clearly explained to all concerned. If the chorionic villi are karyotyped the patient must be told if abnormalities are detected.
4. The informed participating women need to know the steps undertaken. They need reassurance that these steps are not likely to cause harm. We inform all our volunteers that they can exercise their prerogative to refuse to support the training programme.
5. The interest of ancillary medical staff must also be considered. They should be aware of the intent of the programme, so those with reservations can be redeployed, to everyone's mutual benefit.
6. Practice attempts at sampling are best performed with the women anaesthetized, and immediately before surgery. Once the technique is perfected then the second stage—sampling from awake women—can be attempted. We strongly advocate following such a two-stage training programme. Awake women tolerate much less manipulation, and re-

adjustment of technique is often necessary when one changes from sampling in the theatre.

Correct approach and proper selection of women allow a recruitment-rate well in excess of 90 per cent in our department. Interestingly, many women expressed the opinion that they are pleased their pregancies have served a purpose, and were not merely wasted.

10.3 Ultrasound scanning

Most practising obstetricians, particularly the younger generation, will have some background knowledge of ultrasound scanning. This knowledge, however, is usually directed to scanning of mid-trimester or later pregnancies. It is important to stress again that there are certain peculiarities associated with first-trimester scanning, and obstetricians are well-advised to become familiar with these. An example is the mobility of the uterus. The full bladder, the pressure of the scanning probe, and the lithotomy position can all alter the position of the chosen optimum placental site for sampling.

Many obstetricians perform villus sampling with the assistance of an ultrasonographer. This is essentially a very clumsy approach. Co-ordination between the scanner and the operator is never as good as when both these tasks are performed by the operator alone, or by what has been termed 'single operator techique' (Warren *et al.* 1987). A certain amount of practice is required to familiarize oneself with scanning for either trans-abdominal and/or transcervical villus sampling. This again supports the statement that chorion villus sampling is associated with a slightly longer learning curve than for procedures such as mid-trimester amniocentesis.

10.4 Numbers of women required for training

Between 30 and 50 training patients have been proposed as suitable to achieve clinical proficiency. These figures are totally arbitrary: we all learn at different rates. What must be held paramount is the concept that thorough training will reduce the intervention loss-rate for clinical patients. A high fetal loss-rate will also indicate clearly to one's peers the inadequacy of training, and reflects badly on the skills of the operator. Training should be continued until sufficient diagnostic material, between 10 mg and 40 mg, can be obtained with two sampling attempts in over 90 per cent of the time. Only once this standard has been achieved is it time to graduate to awake patients. The move to the clinical situation is made only when the same standard is reached for the second stage as for the first stage of training. As with all intricate procedures, constant practice is essential to maintain the feel of the technique. In a regional diagnostic centre there is seldom any difficulty in achieving this. The situation in other departments may be different. When the number of diagnostic patients falls short, expertise can be maintained by returning to women requesting therapeutic abortion.

10.5 What approach? What implement?

The debate about the approach for villus sampling has now settled into a clear understanding that it is useful to be familiar with both the transabdominal and the transcervical routes. Both approaches have advantages and disadvantages. Which route is chosen as the primary approach is now often a matter of personal preference. In twin pregnancies the placental sites may suggest that a combination of both approaches is appropriate. Some obstetricians, however, may not agree, and feel that the transabdominal or transcervical approach alone is adequate in all circumstances. There is a noticeable drift towards the transabdominal approach in the last few years. It is however true that the transabdominal approach is often easier towards the end of the first trimester, that is at twelve rather than at ten weeks. If cytoculture of the villus material is needed results will not be available to allow early resolution of the situation, thereby defeating the original purpose of villus sampling. The author uses the transcervical route exclusively for all first-trimester sampling. The transabdominal approach is reserved for mid-trimester and latter gestation: an issue with different connotations to that of early prenatal diagnosis.

The choice of the implement used is also a matter of individual preference. It is common sense to suggest that the implement must be one which the operator finds comfortable to use, yet which fulfils the necessary requirements. Selection of implements can be carried out at the training stage. When a suitable implement is found then exclusive use of it will allow development of expertise and understanding of the implement's idiosyncrasies.

10.6 Development of ancillary support services

Molecular biologists and cytogenetic departments will need time to become familiar with use of villus material. Placental tissue obtained at therapeutic abortions can be deployed for laboratory training. When preliminary expertise is achieved then attention can be directed to the smaller volume obtained from villus sampling. For reasons discussed in Chapter 4 many cytogenetic laboratories use direct-preparation techniques as a screen, and provide a final result after karyotyping following cytoculture. If this programme is used, then the appointed time for sampling must be co-ordinated so that results will be available within the first trimester. In our department karyotyping failure is in the region of 2–3 per cent. This figure must be taken into consideration during counselling.

No department is comprehensive for all available diagnostic tests. Departments with particular expertise need to be identified, and their capacity to provide a result must be discussed with the patients. Although cytogeneticists like to process the material as soon after collection as possible, chorionic villi can be transported in culture medium. This is obviously an important issue for the many hospitals without on-site cytogenetic facilities.

10.7 Recruitment of diagnostic patients

Until such time as chorion villus sampling becomes integrated into routine obstetric practice, some organization is needed to ensure an efficient service. Attendance or referrals to the booking clinic are often scheduled for mid-trimester. Although this is considered practical, many women who may need prenatal diagnosis will be denied the benefit of an early test. Patients with a past history of inheritable conditions are usually given prior advice to attend early for their next pregnancies. For the community at large an educational programme to encourage attendance at genetics or pre-pregnancy counselling clinics or their practitioners' surgeries is necessary if there is a family history of likely problems. This will allow pre-pregnancy screenings, such as those for cystic fibrosis or thalassaemia, to be conducted well in advance of a proposed pregnancy. Most women now appreciate that diagnosis for Down syndrome is available if they are thirty-five years of age or older. Educational programmes, such as lectures or written articles, must also be directed at community midwives, paramedical support groups, and, especially, general practitioners, to provide them with the basic knowledge needed to offer informed advice on chorion villus sampling. These health carers can in turn serve the purpose of providing a teaching and screening service to the public.

Early diagnosis, and women's wishes to keep their pregnancies private until normality is established, have given rise to a special problem in certain circumstances. Self-referral or general practitioners' referral for diagnosis may not uncommonly take place before a consultant is selected for antenatal care. Obstetricians performing villus sampling must make it quite clear that referral to them for diagnosis does not automatically imply acceptance for antenatal care. To relieve anxiety in these circumstances arrangements may be made with the women for news of favourable results to be passed on to them directly at the same time as it is passed to the referring doctors.

However, in the event of an adverse result it is the referring doctors who must convey the information to their patients. It is the referring doctors who have the prerogative of arranging the resolution of such a situation; but if no antenatal consultant has yet been appointed it may well seem expedient and appropriate for the obstetrician who has performed the diagnosis to be made responsible for the woman's continued care thereafter.

10.8 Counselling

Guidelines for prenatal diagnosis have been established for over a decade. (Powledge and Fletcher 1979; Clinical Genetics Society 1978). Many women usually have some idea of what they require when they present for diagnosis; but further counselling with particular emphasis on villus sampling is worth while. The following points should be kept in mind:

1. Counselling must be non-directive. Risks and limitations of villus sampling must be clearly explained.
2. Where there are conflicting anxieties, for example with a thirty-nine-year-old woman presenting with her first conception following *in vitro* fertilization, options and alternatives should be explained. They may want to await the outcome of maternal serum indicators screening (Wald and Cuckle 1987) before considering amniocentesis or mid-trimester transabdominal villus sampling.
3. Allow time for couples to discuss the situation between themselves before coming to a decision to proceed with the test. After meticulous explanation they may rightly come to the decision that the risk of abnormality is very low, and diagnosis is not warranted.
4. After sampling, the woman should be advised against coitus within the first few days following the procedure. She should be given clear instructions of what to expect and what constitutes a complication. There must also be clear instructions as to whom to contact if they are worried or when problems arise.
5. A woman should be given some idea when results can be expected. Arrange with her an agreed time and place to relay results. Most women will want to receive adverse results at home, when their partners or family are around for support.
6. When interpretation of results presents difficulty the referring doctor must be notified immediately, and arrangements should be made to convey results to the woman. Many women will want to return to the hospital for further discussion. Mosaicism may need fetal blood

sampling or amniocentesis for verification. The outcome of sex-chromosome abnormalities is not always easy to predict, and necessitates careful counselling.

7. Referring doctors should convey adverse results immediately, and allow time for questions and expression of grief. If therapeutic abortion of pregnancy is appropriate, arrangements must be made as soon as possible.
8. Where possible products of termination should be collected and examined to verify the diagnosis.
9. Opportunity for supportive counselling should be provided after therapeutic abortion. The general practitioner and voluntary support-groups can be deployed to help. Anxieties can extend to subsequent pregnancies. When these women present again in another pregnancy sympathetic support is required throughout any prenatal diagnostic procedures. This support must be continued into the antenatal period, and even after the birth of the child (Lloyd and Lawrence 1985).

10.9 Nursing support

Regional or sub-regional prenatal diagnostic units will have a minimum of two support nurses, usually midwives, who can provide all-round assistance. With training they can co-ordinate the many facets described above to ensure an efficient service. They can participate in counselling, or in scanning, or to co-ordinate the most appropriate time for diagnosis, or to act as a liaison to advise, comfort, and relay results. In addition their role can include auditing, research, and teaching.

At non-referral centres such support may not be available, but this need not detract from the ability to provide a service. Eddy, from a district hospital in Colchester, has described how an efficient villus sampling service can be organized by liaison with a nearby cytogenetic department (1987). Chorionic villi are collected and conveyed in culture medium to the appointed cytogeneticist, who will either perform the karyotyping themselves or use designated centres to provide a more specific diagnosis. Meanwhile the consultant who performs the sampling continues as the contact for the women using the service.

10.10 Costs

The cost-effectiveness of prenatal diagnosis, on both an emotional and a financial level, is convincing (Henderson 1987). Chorion villus sampling takes this effectiveness a step further. Karyotyping of villus material, particularly with the direct-preparation technique, takes less time and need cost no more than that for amniocytes. First-trimester resolution of an unwanted pregnancy in a day-case unit, instead of mid-trimester therapeutic abortion, saves the Health Service anything from £800–£1000 per patient. In a referral centre this saving can easily cover half the annual running cost of a prenatal diagnostic unit.

10.11 Conclusion

The rapid incorporation of chorion villus sampling into routine prenatal diagnosis is a testimony of the growing interest in the technique and the realization of its many advantages for women and doctors alike. It may be new to some of us now; but readers of this text will appreciate that the technique can easily become part of their clinical repertoire, allowing their patients the utility of villus sampling.

References

Clinical Genetics Society. Working Party on Prenatal Diagnosis in Relation to Genetic Counselling (1978). *The provision of services to the prenatal diagnosis of fetal abnormality in the United Kingdom.* Supplement 3. The Society, London.

Eddy, J. W. (1987). Chorion villus sampling for direct chromosomal analysis in a non-teaching hospital. In *Chorion villus sampling,* (ed. D. T. Y. Liu, E. M. Symonds, and M. S. Golbus), pp. 235–44. Chapman and Hall, London.

Henderson, B. J. (1987). The economic efficiency of prenatal screening. In *Chorion villus sampling,* (ed. D. T. Y. Liu, E. M. Symonds, and M. S. Golbus), pp. 245–53. Chapman and Hall, London.

Lloyd, J. and Lawrence, K. M. (1985). Sequelae and support after termination of pregnancy for fetal malformation. *British Medical Journal,* **290**, 907–90.

Powledge, T. M. and Fletcher, J. (1979). Guidelines for the clinical, social and legal issues in prenatal diagnosis. *New England Journal of Medicine,* **300**, 168–72.

Wald, N. J. and Cuckle, H. S. (1987). Recent advances in screening for neural tube defects and Down's syndrome. In *Clinical Obstetrics and Gynaecology. Fetal Diagnosis of Genetic Defects.* (ed. C. H. Rodeck), **1**(3), 649–76. Baillière Tindall, Great Britain.

Warren, R. C., McKenzie, C. F., Kearney, L. and Rodeck, C. H. (1987). Transcervical chorion villus aspiration by a single operator technique. In *Chorion villus sampling,* (ed. D. T. Y. Liu, E. M. Symonds, and M. S. Golbus), pp. 65–72. Chapman and Hall, London.

Index